Invasive Species: A Very Short Introduction

VERY SHORT INTRODUCTIONS are for anyone wanting a stimulating and accessible way into a new subject. They are written by experts, and have been translated into more than 45 different languages.

The series began in 1995, and now covers a wide variety of topics in every discipline. The VSI library currently contains over 700 volumes—a Very Short Introduction to everything from Psychology and Philosophy of Science to American History and Relativity—and continues to grow in every subject area.

Very Short Introductions available now:

ABOLITIONISM Richard S. Newman
THE ABRAHAMIC RELIGIONS Charles L. Cohen
ACCOUNTING Christopher Nobes
ADDICTION Keith Humphreys
ADOLESCENCE Peter K. Smith
THEODOR W. ADORNO Andrew Bowie
ADVERTISING Winston Fletcher
AERIAL WARFARE Frank Ledwidge
AESTHETICS Bence Nanay
AFRICAN AMERICAN HISTORY Jonathan Scott Holloway
AFRICAN AMERICAN RELIGION Eddie S. Glaude Jr
AFRICAN HISTORY John Parker and Richard Rathbone
AFRICAN POLITICS Ian Taylor
AFRICAN RELIGIONS Jacob K. Olupona
AGEING Nancy A. Pachana
AGNOSTICISM Robin Le Poidevin
AGRICULTURE Paul Brassley and Richard Soffe
ALEXANDER THE GREAT Hugh Bowden
ALGEBRA Peter M. Higgins
AMERICAN BUSINESS HISTORY Walter A. Friedman
AMERICAN CULTURAL HISTORY Eric Avila
AMERICAN FOREIGN RELATIONS Andrew Preston
AMERICAN HISTORY Paul S. Boyer
AMERICAN IMMIGRATION David A. Gerber
AMERICAN INTELLECTUAL HISTORY Jennifer Ratner-Rosenhagen
THE AMERICAN JUDICIAL SYSTEM Charles L. Zelden
AMERICAN LEGAL HISTORY G. Edward White
AMERICAN MILITARY HISTORY Joseph T. Glatthaar
AMERICAN NAVAL HISTORY Craig L. Symonds
AMERICAN POETRY David Caplan
AMERICAN POLITICAL HISTORY Donald Critchlow
AMERICAN POLITICAL PARTIES AND ELECTIONS L. Sandy Maisel
AMERICAN POLITICS Richard M. Valelly
THE AMERICAN PRESIDENCY Charles O. Jones
THE AMERICAN REVOLUTION Robert J. Allison
AMERICAN SLAVERY Heather Andrea Williams
THE AMERICAN SOUTH Charles Reagan Wilson
THE AMERICAN WEST Stephen Aron
AMERICAN WOMEN'S HISTORY Susan Ware
AMPHIBIANS T. S. Kemp
ANAESTHESIA Aidan O'Donnell

ANALYTIC PHILOSOPHY
Michael Beaney
ANARCHISM Alex Prichard
ANCIENT ASSYRIA Karen Radner
ANCIENT EGYPT Ian Shaw
ANCIENT EGYPTIAN ART AND ARCHITECTURE Christina Riggs
ANCIENT GREECE Paul Cartledge
ANCIENT GREEK AND ROMAN SCIENCE Liba Taub
THE ANCIENT NEAR EAST
Amanda H. Podany
ANCIENT PHILOSOPHY Julia Annas
ANCIENT WARFARE
Harry Sidebottom
ANGELS David Albert Jones
ANGLICANISM Mark Chapman
THE ANGLO-SAXON AGE John Blair
ANIMAL BEHAVIOUR
Tristram D. Wyatt
THE ANIMAL KINGDOM
Peter Holland
ANIMAL RIGHTS David DeGrazia
ANSELM Thomas Williams
THE ANTARCTIC Klaus Dodds
ANTHROPOCENE Erle C. Ellis
ANTISEMITISM Steven Beller
ANXIETY Daniel Freeman and Jason Freeman
THE APOCRYPHAL GOSPELS
Paul Foster
APPLIED MATHEMATICS
Alain Goriely
THOMAS AQUINAS Fergus Kerr
ARBITRATION Thomas Schultz and Thomas Grant
ARCHAEOLOGY Paul Bahn
ARCHITECTURE Andrew Ballantyne
THE ARCTIC Klaus Dodds and Jamie Woodward
HANNAH ARENDT Dana Villa
ARISTOCRACY William Doyle
ARISTOTLE Jonathan Barnes
ART HISTORY Dana Arnold
ART THEORY Cynthia Freeland
ARTIFICIAL INTELLIGENCE
Margaret A. Boden
ASIAN AMERICAN HISTORY
Madeline Y. Hsu
ASTROBIOLOGY David C. Catling
ASTROPHYSICS James Binney
ATHEISM Julian Baggini
THE ATMOSPHERE Paul I. Palmer
AUGUSTINE Henry Chadwick
JANE AUSTEN Tom Keymer
AUSTRALIA Kenneth Morgan
AUTISM Uta Frith
AUTOBIOGRAPHY Laura Marcus
THE AVANT GARDE David Cottington
THE AZTECS Davíd Carrasco
BABYLONIA Trevor Bryce
BACTERIA Sebastian G. B. Amyes
BANKING John Goddard and John O. S. Wilson
BARTHES Jonathan Culler
THE BEATS David Sterritt
BEAUTY Roger Scruton
LUDWIG VAN BEETHOVEN
Mark Evan Bonds
BEHAVIOURAL ECONOMICS
Michelle Baddeley
BESTSELLERS John Sutherland
THE BIBLE John Riches
BIBLICAL ARCHAEOLOGY
Eric H. Cline
BIG DATA Dawn E. Holmes
BIOCHEMISTRY Mark Lorch
BIODIVERSITY CONSERVATION
David Macdonald
BIOGEOGRAPHY Mark V. Lomolino
BIOGRAPHY Hermione Lee
BIOMETRICS Michael Fairhurst
ELIZABETH BISHOP Jonathan F. S. Post
BLACK HOLES Katherine Blundell
BLASPHEMY Yvonne Sherwood
BLOOD Chris Cooper
THE BLUES Elijah Wald
THE BODY Chris Shilling
THE BOHEMIANS David Weir
NIELS BOHR J. L. Heilbron
THE BOOK OF COMMON PRAYER
Brian Cummings
THE BOOK OF MORMON
Terryl Givens
BORDERS Alexander C. Diener and Joshua Hagen
THE BRAIN Michael O'Shea
BRANDING Robert Jones
THE BRICS Andrew F. Cooper
BRITISH CINEMA Charles Barr

THE BRITISH CONSTITUTION Martin Loughlin
THE BRITISH EMPIRE Ashley Jackson
BRITISH POLITICS Tony Wright
BUDDHA Michael Carrithers
BUDDHISM Damien Keown
BUDDHIST ETHICS Damien Keown
BYZANTIUM Peter Sarris
CALVINISM Jon Balserak
ALBERT CAMUS Oliver Gloag
CANADA Donald Wright
CANCER Nicholas James
CAPITALISM James Fulcher
CATHOLICISM Gerald O'Collins
CAUSATION Stephen Mumford and Rani Lill Anjum
THE CELL Terence Allen and Graham Cowling
THE CELTS Barry Cunliffe
CHAOS Leonard Smith
GEOFFREY CHAUCER David Wallace
CHEMISTRY Peter Atkins
CHILD PSYCHOLOGY Usha Goswami
CHILDREN'S LITERATURE Kimberley Reynolds
CHINESE LITERATURE Sabina Knight
CHOICE THEORY Michael Allingham
CHRISTIAN ART Beth Williamson
CHRISTIAN ETHICS D. Stephen Long
CHRISTIANITY Linda Woodhead
CIRCADIAN RHYTHMS Russell Foster and Leon Kreitzman
CITIZENSHIP Richard Bellamy
CITY PLANNING Carl Abbott
CIVIL ENGINEERING David Muir Wood
THE CIVIL RIGHTS MOVEMENT Thomas C. Holt
CLASSICAL LITERATURE William Allan
CLASSICAL MYTHOLOGY Helen Morales
CLASSICS Mary Beard and John Henderson
CLAUSEWITZ Michael Howard
CLIMATE Mark Maslin
CLIMATE CHANGE Mark Maslin
CLINICAL PSYCHOLOGY Susan Llewelyn and Katie Aafjes-van Doorn
COGNITIVE BEHAVIOURAL THERAPY Freda McManus
COGNITIVE NEUROSCIENCE Richard Passingham
THE COLD WAR Robert J. McMahon
COLONIAL AMERICA Alan Taylor
COLONIAL LATIN AMERICAN LITERATURE Rolena Adorno
COMBINATORICS Robin Wilson
COMEDY Matthew Bevis
COMMUNISM Leslie Holmes
COMPARATIVE LITERATURE Ben Hutchinson
COMPETITION AND ANTITRUST LAW Ariel Ezrachi
COMPLEXITY John H. Holland
THE COMPUTER Darrel Ince
COMPUTER SCIENCE Subrata Dasgupta
CONCENTRATION CAMPS Dan Stone
CONDENSED MATTER PHYSICS Ross H. McKenzie
CONFUCIANISM Daniel K. Gardner
THE CONQUISTADORS Matthew Restall and Felipe Fernández-Armesto
CONSCIENCE Paul Strohm
CONSCIOUSNESS Susan Blackmore
CONTEMPORARY ART Julian Stallabrass
CONTEMPORARY FICTION Robert Eaglestone
CONTINENTAL PHILOSOPHY Simon Critchley
COPERNICUS Owen Gingerich
CORAL REEFS Charles Sheppard
CORPORATE SOCIAL RESPONSIBILITY Jeremy Moon
CORRUPTION Leslie Holmes
COSMOLOGY Peter Coles
COUNTRY MUSIC Richard Carlin
CREATIVITY Vlad Glăveanu
CRIME FICTION Richard Bradford
CRIMINAL JUSTICE Julian V. Roberts
CRIMINOLOGY Tim Newburn
CRITICAL THEORY Stephen Eric Bronner
THE CRUSADES Christopher Tyerman

CRYPTOGRAPHY Fred Piper and Sean Murphy
CRYSTALLOGRAPHY A. M. Glazer
THE CULTURAL REVOLUTION Richard Curt Kraus
DADA AND SURREALISM David Hopkins
DANTE Peter Hainsworth and David Robey
DARWIN Jonathan Howard
THE DEAD SEA SCROLLS Timothy H. Lim
DECADENCE David Weir
DECOLONIZATION Dane Kennedy
DEMENTIA Kathleen Taylor
DEMOCRACY Naomi Zack
DEMOGRAPHY Sarah Harper
DEPRESSION Jan Scott and Mary Jane Tacchi
DERRIDA Simon Glendinning
DESCARTES Tom Sorell
DESERTS Nick Middleton
DESIGN John Heskett
DEVELOPMENT Ian Goldin
DEVELOPMENTAL BIOLOGY Lewis Wolpert
THE DEVIL Darren Oldridge
DIASPORA Kevin Kenny
CHARLES DICKENS Jenny Hartley
DICTIONARIES Lynda Mugglestone
DINOSAURS David Norman
DIPLOMATIC HISTORY Joseph M. Siracusa
DOCUMENTARY FILM Patricia Aufderheide
DREAMING J. Allan Hobson
DRUGS Les Iversen
DRUIDS Barry Cunliffe
DYNASTY Jeroen Duindam
DYSLEXIA Margaret J. Snowling
EARLY MUSIC Thomas Forrest Kelly
THE EARTH Martin Redfern
EARTH SYSTEM SCIENCE Tim Lenton
ECOLOGY Jaboury Ghazoul
ECONOMICS Partha Dasgupta
EDUCATION Gary Thomas
EGYPTIAN MYTH Geraldine Pinch
EIGHTEENTH-CENTURY BRITAIN Paul Langford
THE ELEMENTS Philip Ball
EMOTION Dylan Evans
EMPIRE Stephen Howe
EMPLOYMENT LAW David Cabrelli
ENERGY SYSTEMS Nick Jenkins
ENGELS Terrell Carver
ENGINEERING David Blockley
THE ENGLISH LANGUAGE Simon Horobin
ENGLISH LITERATURE Jonathan Bate
THE ENLIGHTENMENT John Robertson
ENTREPRENEURSHIP Paul Westhead and Mike Wright
ENVIRONMENTAL ECONOMICS Stephen Smith
ENVIRONMENTAL ETHICS Robin Attfield
ENVIRONMENTAL LAW Elizabeth Fisher
ENVIRONMENTAL POLITICS Andrew Dobson
ENZYMES Paul Engel
EPICUREANISM Catherine Wilson
EPIDEMIOLOGY Rodolfo Saracci
ETHICS Simon Blackburn
ETHNOMUSICOLOGY Timothy Rice
THE ETRUSCANS Christopher Smith
EUGENICS Philippa Levine
THE EUROPEAN UNION Simon Usherwood and John Pinder
EUROPEAN UNION LAW Anthony Arnull
EVANGELICALISM John G. Stackhouse Jr.
EVIL Luke Russell
EVOLUTION Brian and Deborah Charlesworth
EXISTENTIALISM Thomas Flynn
EXPLORATION Stewart A. Weaver
EXTINCTION Paul B. Wignall
THE EYE Michael Land
FAIRY TALE Marina Warner
FAMILY LAW Jonathan Herring
MICHAEL FARADAY Frank A. J. L. James
FASCISM Kevin Passmore
FASHION Rebecca Arnold

FEDERALISM Mark J. Rozell and Clyde Wilcox
FEMINISM Margaret Walters
FILM Michael Wood
FILM MUSIC Kathryn Kalinak
FILM NOIR James Naremore
FIRE Andrew C. Scott
THE FIRST WORLD WAR Michael Howard
FLUID MECHANICS Eric Lauga
FOLK MUSIC Mark Slobin
FOOD John Krebs
FORENSIC PSYCHOLOGY David Canter
FORENSIC SCIENCE Jim Fraser
FORESTS Jaboury Ghazoul
FOSSILS Keith Thomson
FOUCAULT Gary Gutting
THE FOUNDING FATHERS R. B. Bernstein
FRACTALS Kenneth Falconer
FREE SPEECH Nigel Warburton
FREE WILL Thomas Pink
FREEMASONRY Andreas Önnerfors
FRENCH LITERATURE John D. Lyons
FRENCH PHILOSOPHY Stephen Gaukroger and Knox Peden
THE FRENCH REVOLUTION William Doyle
FREUD Anthony Storr
FUNDAMENTALISM Malise Ruthven
FUNGI Nicholas P. Money
THE FUTURE Jennifer M. Gidley
GALAXIES John Gribbin
GALILEO Stillman Drake
GAME THEORY Ken Binmore
GANDHI Bhikhu Parekh
GARDEN HISTORY Gordon Campbell
GENES Jonathan Slack
GENIUS Andrew Robinson
GENOMICS John Archibald
GEOGRAPHY John Matthews and David Herbert
GEOLOGY Jan Zalasiewicz
GEOMETRY Maciej Dunajski
GEOPHYSICS William Lowrie
GEOPOLITICS Klaus Dodds
GERMAN LITERATURE Nicholas Boyle
GERMAN PHILOSOPHY Andrew Bowie
THE GHETTO Bryan Cheyette
GLACIATION David J. A. Evans
GLOBAL CATASTROPHES Bill McGuire
GLOBAL ECONOMIC HISTORY Robert C. Allen
GLOBAL ISLAM Nile Green
GLOBALIZATION Manfred B. Steger
GOD John Bowker
GÖDEL'S THEOREM A. W. Moore
GOETHE Ritchie Robertson
THE GOTHIC Nick Groom
GOVERNANCE Mark Bevir
GRAVITY Timothy Clifton
THE GREAT DEPRESSION AND THE NEW DEAL Eric Rauchway
HABEAS CORPUS Amanda L. Tyler
HABERMAS James Gordon Finlayson
THE HABSBURG EMPIRE Martyn Rady
HAPPINESS Daniel M. Haybron
THE HARLEM RENAISSANCE Cheryl A. Wall
THE HEBREW BIBLE AS LITERATURE Tod Linafelt
HEGEL Peter Singer
HEIDEGGER Michael Inwood
THE HELLENISTIC AGE Peter Thonemann
HEREDITY John Waller
HERMENEUTICS Jens Zimmermann
HERODOTUS Jennifer T. Roberts
HIEROGLYPHS Penelope Wilson
HINDUISM Kim Knott
HISTORY John H. Arnold
THE HISTORY OF ASTRONOMY Michael Hoskin
THE HISTORY OF CHEMISTRY William H. Brock
THE HISTORY OF CHILDHOOD James Marten
THE HISTORY OF CINEMA Geoffrey Nowell-Smith
THE HISTORY OF COMPUTING Doron Swade
THE HISTORY OF EMOTIONS Thomas Dixon
THE HISTORY OF LIFE Michael Benton
THE HISTORY OF MATHEMATICS Jacqueline Stedall

THE HISTORY OF MEDICINE William Bynum
THE HISTORY OF PHYSICS J. L. Heilbron
THE HISTORY OF POLITICAL THOUGHT Richard Whatmore
THE HISTORY OF TIME Leofranc Holford-Strevens
HIV AND AIDS Alan Whiteside
HOBBES Richard Tuck
HOLLYWOOD Peter Decherney
THE HOLY ROMAN EMPIRE Joachim Whaley
HOME Michael Allen Fox
HOMER Barbara Graziosi
HORMONES Martin Luck
HORROR Darryl Jones
HUMAN ANATOMY Leslie Klenerman
HUMAN EVOLUTION Bernard Wood
HUMAN PHYSIOLOGY Jamie A. Davies
HUMAN RESOURCE MANAGEMENT Adrian Wilkinson
HUMAN RIGHTS Andrew Clapham
HUMANISM Stephen Law
HUME James A. Harris
HUMOUR Noël Carroll
IBN SĪNĀ Peter Adamson
THE ICE AGE Jamie Woodward
IDENTITY Florian Coulmas
IDEOLOGY Michael Freeden
THE IMMUNE SYSTEM Paul Klenerman
INDIAN CINEMA Ashish Rajadhyaksha
INDIAN PHILOSOPHY Sue Hamilton
THE INDUSTRIAL REVOLUTION Robert C. Allen
INFECTIOUS DISEASE Marta L. Wayne and Benjamin M. Bolker
INFINITY Ian Stewart
INFORMATION Luciano Floridi
INNOVATION Mark Dodgson and David Gann
INTELLECTUAL PROPERTY Siva Vaidhyanathan
INTELLIGENCE Ian J. Deary
INTERNATIONAL LAW Vaughan Lowe
INTERNATIONAL MIGRATION Khalid Koser
INTERNATIONAL RELATIONS Christian Reus-Smit
INTERNATIONAL SECURITY Christopher S. Browning
INSECTS Simon Leather
INVASIVE SPECIES Julie Lockwood and Dustin Welbourne
IRAN Ali M. Ansari
ISLAM Malise Ruthven
ISLAMIC HISTORY Adam Silverstein
ISLAMIC LAW Mashood A. Baderin
ISOTOPES Rob Ellam
ITALIAN LITERATURE Peter Hainsworth and David Robey
HENRY JAMES Susan L. Mizruchi
JAPANESE LITERATURE Alan Tansman
JESUS Richard Bauckham
JEWISH HISTORY David N. Myers
JEWISH LITERATURE Ilan Stavans
JOURNALISM Ian Hargreaves
JAMES JOYCE Colin MacCabe
JUDAISM Norman Solomon
JUNG Anthony Stevens
THE JURY Renée Lettow Lerner
KABBALAH Joseph Dan
KAFKA Ritchie Robertson
KANT Roger Scruton
KEYNES Robert Skidelsky
KIERKEGAARD Patrick Gardiner
KNOWLEDGE Jennifer Nagel
THE KORAN Michael Cook
KOREA Michael J. Seth
LAKES Warwick F. Vincent
LANDSCAPE ARCHITECTURE Ian H. Thompson
LANDSCAPES AND GEOMORPHOLOGY Andrew Goudie and Heather Viles
LANGUAGES Stephen R. Anderson
LATE ANTIQUITY Gillian Clark
LAW Raymond Wacks
THE LAWS OF THERMODYNAMICS Peter Atkins
LEADERSHIP Keith Grint
LEARNING Mark Haselgrove
LEIBNIZ Maria Rosa Antognazza
C. S. LEWIS James Como
LIBERALISM Michael Freeden

LIGHT Ian Walmsley
LINCOLN Allen C. Guelzo
LINGUISTICS Peter Matthews
LITERARY THEORY Jonathan Culler
LOCKE John Dunn
LOGIC Graham Priest
LOVE Ronald de Sousa
MARTIN LUTHER Scott H. Hendrix
MACHIAVELLI Quentin Skinner
MADNESS Andrew Scull
MAGIC Owen Davies
MAGNA CARTA Nicholas Vincent
MAGNETISM Stephen Blundell
MALTHUS Donald Winch
MAMMALS T. S. Kemp
MANAGEMENT John Hendry
NELSON MANDELA Elleke Boehmer
MAO Delia Davin
MARINE BIOLOGY Philip V. Mladenov
MARKETING Kenneth Le Meunier-FitzHugh
THE MARQUIS DE SADE John Phillips
MARTYRDOM Jolyon Mitchell
MARX Peter Singer
MATERIALS Christopher Hall
MATHEMATICAL ANALYSIS Richard Earl
MATHEMATICAL FINANCE Mark H. A. Davis
MATHEMATICS Timothy Gowers
MATTER Geoff Cottrell
THE MAYA Matthew Restall and Amara Solari
THE MEANING OF LIFE Terry Eagleton
MEASUREMENT David Hand
MEDICAL ETHICS Michael Dunn and Tony Hope
MEDICAL LAW Charles Foster
MEDIEVAL BRITAIN John Gillingham and Ralph A. Griffiths
MEDIEVAL LITERATURE Elaine Treharne
MEDIEVAL PHILOSOPHY John Marenbon
MEMORY Jonathan K. Foster
METAPHYSICS Stephen Mumford
METHODISM William J. Abraham
THE MEXICAN REVOLUTION Alan Knight
MICROBIOLOGY Nicholas P. Money
MICROBIOMES Angela E. Douglas
MICROECONOMICS Avinash Dixit
MICROSCOPY Terence Allen
THE MIDDLE AGES Miri Rubin
MILITARY JUSTICE Eugene R. Fidell
MILITARY STRATEGY Antulio J. Echevarria II
JOHN STUART MILL Gregory Claeys
MINERALS David Vaughan
MIRACLES Yujin Nagasawa
MODERN ARCHITECTURE Adam Sharr
MODERN ART David Cottington
MODERN BRAZIL Anthony W. Pereira
MODERN CHINA Rana Mitter
MODERN DRAMA Kirsten E. Shepherd-Barr
MODERN FRANCE Vanessa R. Schwartz
MODERN INDIA Craig Jeffrey
MODERN IRELAND Senia Pašeta
MODERN ITALY Anna Cento Bull
MODERN JAPAN Christopher Goto-Jones
MODERN LATIN AMERICAN LITERATURE Roberto González Echevarría
MODERN WAR Richard English
MODERNISM Christopher Butler
MOLECULAR BIOLOGY Aysha Divan and Janice A. Royds
MOLECULES Philip Ball
MONASTICISM Stephen J. Davis
THE MONGOLS Morris Rossabi
MONTAIGNE William M. Hamlin
MOONS David A. Rothery
MORMONISM Richard Lyman Bushman
MOUNTAINS Martin F. Price
MUHAMMAD Jonathan A. C. Brown
MULTICULTURALISM Ali Rattansi
MULTILINGUALISM John C. Maher
MUSIC Nicholas Cook
MUSIC AND TECHNOLOGY Mark Katz
MYTH Robert A. Segal
NANOTECHNOLOGY Philip Moriarty
NAPOLEON David A. Bell
THE NAPOLEONIC WARS Mike Rapport
NATIONALISM Steven Grosby

NATIVE AMERICAN LITERATURE Sean Teuton
NAVIGATION Jim Bennett
NAZI GERMANY Jane Caplan
NEGOTIATION Carrie Menkel-Meadow
NEOLIBERALISM Manfred B. Steger and Ravi K. Roy
NETWORKS Guido Caldarelli and Michele Catanzaro
THE NEW TESTAMENT Luke Timothy Johnson
THE NEW TESTAMENT AS LITERATURE Kyle Keefer
NEWTON Robert Iliffe
NIETZSCHE Michael Tanner
NINETEENTH-CENTURY BRITAIN Christopher Harvie and H. C. G. Matthew
THE NORMAN CONQUEST George Garnett
NORTH AMERICAN INDIANS Theda Perdue and Michael D. Green
NORTHERN IRELAND Marc Mulholland
NOTHING Frank Close
NUCLEAR PHYSICS Frank Close
NUCLEAR POWER Maxwell Irvine
NUCLEAR WEAPONS Joseph M. Siracusa
NUMBER THEORY Robin Wilson
NUMBERS Peter M. Higgins
NUTRITION David A. Bender
OBJECTIVITY Stephen Gaukroger
OBSERVATIONAL ASTRONOMY Geoff Cottrell
OCEANS Dorrik Stow
THE OLD TESTAMENT Michael D. Coogan
THE ORCHESTRA D. Kern Holoman
ORGANIC CHEMISTRY Graham Patrick
ORGANIZATIONS Mary Jo Hatch
ORGANIZED CRIME Georgios A. Antonopoulos and Georgios Papanicolaou
ORTHODOX CHRISTIANITY A. Edward Siecienski
OVID Llewelyn Morgan
PAGANISM Owen Davies
PAKISTAN Pippa Virdee
THE PALESTINIAN-ISRAELI CONFLICT Martin Bunton
PANDEMICS Christian W. McMillen
PARTICLE PHYSICS Frank Close
PAUL E. P. Sanders
IVAN PAVLOV Daniel P. Todes
PEACE Oliver P. Richmond
PENTECOSTALISM William K. Kay
PERCEPTION Brian Rogers
THE PERIODIC TABLE Eric R. Scerri
PHILOSOPHICAL METHOD Timothy Williamson
PHILOSOPHY Edward Craig
PHILOSOPHY IN THE ISLAMIC WORLD Peter Adamson
PHILOSOPHY OF BIOLOGY Samir Okasha
PHILOSOPHY OF LAW Raymond Wacks
PHILOSOPHY OF MIND Barbara Gail Montero
PHILOSOPHY OF PHYSICS David Wallace
PHILOSOPHY OF SCIENCE Samir Okasha
PHILOSOPHY OF RELIGION Tim Bayne
PHOTOGRAPHY Steve Edwards
PHYSICAL CHEMISTRY Peter Atkins
PHYSICS Sidney Perkowitz
PILGRIMAGE Ian Reader
PLAGUE Paul Slack
PLANETARY SYSTEMS Raymond T. Pierrehumbert
PLANETS David A. Rothery
PLANTS Timothy Walker
PLATE TECTONICS Peter Molnar
PLATO Julia Annas
POETRY Bernard O'Donoghue
POLITICAL PHILOSOPHY David Miller
POLITICS Kenneth Minogue
POLYGAMY Sarah M. S. Pearsall
POPULISM Cas Mudde and Cristóbal Rovira Kaltwasser
POSTCOLONIALISM Robert J. C. Young
POSTMODERNISM Christopher Butler
POSTSTRUCTURALISM Catherine Belsey
POVERTY Philip N. Jefferson
PREHISTORY Chris Gosden

PRESOCRATIC PHILOSOPHY Catherine Osborne
PRIVACY Raymond Wacks
PROBABILITY John Haigh
PROGRESSIVISM Walter Nugent
PROHIBITION W. J. Rorabaugh
PROJECTS Andrew Davies
PROTESTANTISM Mark A. Noll
PSEUDOSCIENCE Michael D. Gordin.
PSYCHIATRY Tom Burns
PSYCHOANALYSIS Daniel Pick
PSYCHOLOGY Gillian Butler and Freda McManus
PSYCHOLOGY OF MUSIC Elizabeth Hellmuth Margulis
PSYCHOPATHY Essi Viding
PSYCHOTHERAPY Tom Burns and Eva Burns-Lundgren
PUBLIC ADMINISTRATION Stella Z. Theodoulou and Ravi K. Roy
PUBLIC HEALTH Virginia Berridge
PURITANISM Francis J. Bremer
THE QUAKERS Pink Dandelion
QUANTUM THEORY John Polkinghorne
RACISM Ali Rattansi
RADIOACTIVITY Claudio Tuniz
RASTAFARI Ennis B. Edmonds
READING Belinda Jack
THE REAGAN REVOLUTION Gil Troy
REALITY Jan Westerhoff
RECONSTRUCTION Allen C. Guelzo
THE REFORMATION Peter Marshall
REFUGEES Gil Loescher
RELATIVITY Russell Stannard
RELIGION Thomas A. Tweed
RELIGION IN AMERICA Timothy Beal
THE RENAISSANCE Jerry Brotton
RENAISSANCE ART Geraldine A. Johnson
RENEWABLE ENERGY Nick Jelley
REPTILES T. S. Kemp
REVOLUTIONS Jack A. Goldstone
RHETORIC Richard Toye
RISK Baruch Fischhoff and John Kadvany
RITUAL Barry Stephenson
RIVERS Nick Middleton
ROBOTICS Alan Winfield
ROCKS Jan Zalasiewicz
ROMAN BRITAIN Peter Salway
THE ROMAN EMPIRE Christopher Kelly
THE ROMAN REPUBLIC David M. Gwynn
ROMANTICISM Michael Ferber
ROUSSEAU Robert Wokler
RUSSELL A. C. Grayling
THE RUSSIAN ECONOMY Richard Connolly
RUSSIAN HISTORY Geoffrey Hosking
RUSSIAN LITERATURE Catriona Kelly
THE RUSSIAN REVOLUTION S. A. Smith
SAINTS Simon Yarrow
SAMURAI Michael Wert
SAVANNAS Peter A. Furley
SCEPTICISM Duncan Pritchard
SCHIZOPHRENIA Chris Frith and Eve Johnstone
SCHOPENHAUER Christopher Janaway
SCIENCE AND RELIGION Thomas Dixon and Adam R. Shapiro
SCIENCE FICTION David Seed
THE SCIENTIFIC REVOLUTION Lawrence M. Principe
SCOTLAND Rab Houston
SECULARISM Andrew Copson
SEXUAL SELECTION Marlene Zuk and Leigh W. Simmons
SEXUALITY Véronique Mottier
WILLIAM SHAKESPEARE Stanley Wells
SHAKESPEARE'S COMEDIES Bart van Es
SHAKESPEARE'S SONNETS AND POEMS Jonathan F. S. Post
SHAKESPEARE'S TRAGEDIES Stanley Wells
GEORGE BERNARD SHAW Christopher Wixson
MARY SHELLEY Charlotte Gordon
THE SHORT STORY Andrew Kahn
SIKHISM Eleanor Nesbitt
SILENT FILM Donna Kornhaber
THE SILK ROAD James A. Millward
SLANG Jonathon Green
SLEEP Steven W. Lockley and Russell G. Foster
SMELL Matthew Cobb
ADAM SMITH Christopher J. Berry

SOCIAL AND CULTURAL ANTHROPOLOGY John Monaghan and Peter Just
SOCIAL PSYCHOLOGY Richard J. Crisp
SOCIAL WORK Sally Holland and Jonathan Scourfield
SOCIALISM Michael Newman
SOCIOLINGUISTICS John Edwards
SOCIOLOGY Steve Bruce
SOCRATES C. C. W. Taylor
SOFT MATTER Tom McLeish
SOUND Mike Goldsmith
SOUTHEAST ASIA James R. Rush
THE SOVIET UNION Stephen Lovell
THE SPANISH CIVIL WAR Helen Graham
SPANISH LITERATURE Jo Labanyi
THE SPARTANS Andrew J. Bayliss
SPINOZA Roger Scruton
SPIRITUALITY Philip Sheldrake
SPORT Mike Cronin
STARS Andrew King
STATISTICS David J. Hand
STEM CELLS Jonathan Slack
STOICISM Brad Inwood
STRUCTURAL ENGINEERING David Blockley
STUART BRITAIN John Morrill
SUBURBS Carl Abbott
THE SUN Philip Judge
SUPERCONDUCTIVITY Stephen Blundell
SUPERSTITION Stuart Vyse
SYMMETRY Ian Stewart
SYNAESTHESIA Julia Simner
SYNTHETIC BIOLOGY Jamie A. Davies
SYSTEMS BIOLOGY Eberhard O. Voit
TAXATION Stephen Smith
TEETH Peter S. Ungar
TERRORISM Charles Townshend
THEATRE Marvin Carlson
THEOLOGY David F. Ford
THINKING AND REASONING Jonathan St B. T. Evans
THOUGHT Tim Bayne
TIBETAN BUDDHISM Matthew T. Kapstein
TIDES David George Bowers and Emyr Martyn Roberts
TIME Jenann Ismael
TOCQUEVILLE Harvey C. Mansfield
LEO TOLSTOY Liza Knapp
TOPOLOGY Richard Earl
TRAGEDY Adrian Poole
TRANSLATION Matthew Reynolds
THE TREATY OF VERSAILLES Michael S. Neiberg
TRIGONOMETRY Glen Van Brummelen
THE TROJAN WAR Eric H. Cline
TRUST Katherine Hawley
THE TUDORS John Guy
TWENTIETH-CENTURY BRITAIN Kenneth O. Morgan
TYPOGRAPHY Paul Luna
THE UNITED NATIONS Jussi M. Hanhimäki
UNIVERSITIES AND COLLEGES David Palfreyman and Paul Temple
THE U.S. CIVIL WAR Louis P. Masur
THE U.S. CONGRESS Donald A. Ritchie
THE U.S. CONSTITUTION David J. Bodenhamer
THE U.S. SUPREME COURT Linda Greenhouse
UTILITARIANISM Katarzyna de Lazari-Radek and Peter Singer
UTOPIANISM Lyman Tower Sargent
VATICAN II Shaun Blanchard and Stephen Bullivant
VETERINARY SCIENCE James Yeates
THE VIKINGS Julian D. Richards
VIOLENCE Philip Dwyer
THE VIRGIN MARY Mary Joan Winn Leith
THE VIRTUES Craig A. Boyd and Kevin Timpe
VIRUSES Dorothy H. Crawford
VOLCANOES Michael J. Branney and Jan Zalasiewicz
VOLTAIRE Nicholas Cronk
WAR AND RELIGION Jolyon Mitchell and Joshua Rey
WAR AND TECHNOLOGY Alex Roland
WATER John Finney
WAVES Mike Goldsmith
WEATHER Storm Dunlop
THE WELFARE STATE David Garland
WITCHCRAFT Malcolm Gaskill
WITTGENSTEIN A. C. Grayling

WORK Stephen Fineman
WORLD MUSIC Philip Bohlman
WORLD MYTHOLOGY David Leeming
THE WORLD TRADE ORGANIZATION Amrita Narlikar
WORLD WAR II Gerhard L. Weinberg
WRITING AND SCRIPT Andrew Robinson
ZIONISM Michael Stanislawski
ÉMILE ZOLA Brian Nelson

Available soon:

IMAGINATION Jennifer Gosetti-Ferencei
THE VICTORIANS Martin Hewitt

For more information visit our website
www.oup.com/vsi/

Julie L. Lockwood and Dustin J. Welbourne

INVASIVE SPECIES

A Very Short Introduction

Great Clarendon Street, Oxford, OX2 6DP,
United Kingdom

Oxford University Press is a department of the University of Oxford.
It furthers the University's objective of excellence in research, scholarship,
and education by publishing worldwide. Oxford is a registered trade mark of
Oxford University Press in the UK and in certain other countries

Published in the United States of America by Oxford University Press
198 Madison Avenue, New York, NY 10016, United States of America

British Library Cataloguing in Publication Data
Data available

Library of Congress Control Number: 2023932755

ISBN 978–0–19–881828–1

Printed and bound by
CPI Group (UK) Ltd, Croydon, CR0 4YY

Contents

Preface

If humans suddenly vanished from Earth and aliens arrived thousands or even a million years into the future, they would still know we once existed. Our existence would be betrayed by several indelible signatures, perhaps most notably would be the distributions of plants and animals around the globe. Put yourself in their shoes, or whatever it is they wear. After taking an inventory of organisms present in ecosystems—plants, animals, fungi, etc.—they would ask an obvious follow-up question: Why is this species here, and not over there? And they would quickly run into a puzzle.

Prior to some point in time—for our sake let's say 1500 CE—the distributions of organisms could largely be explained by reference to evolution, plate tectonics, and the incremental steps in biological dispersal. But after this point, those explanations no longer hold. Some organisms that were only ever in one area of the planet, say south-eastern Asia or Europe, suddenly appear on other continents, thousands of kilometres from where they evolved. And locations, Australia or New Zealand for instance, isolated for millions of years and home to unique species that occur nowhere else, spontaneously harbour cats and rats.

To make sense of this pivotal change in Earth's history, the aliens would have to conclude that there was a globally distributed species that, for some reason, moved a non-trivial number of species from one area to another. In fact, the dominant species that these aliens might find in those future ecosystems may very well be the species that we have moved. Would these aliens wonder why we did this? Why would a global species deliberately or otherwise allow the biosphere to be reshuffled in this way?

Now, this alien thought experiment might seem a little frivolous, but it highlights a rather serious point: Invasive species have forever, and unequivocally, altered the evolutionary future of Earth. Furthermore, the thought experiment has us wondering whether those aliens would recognize invasive species in the same way we do. Are invasive species universal? Or, to put that question another way, given the development of a civilization, are invasive species inevitable? This is a great question because, even though answering it is perhaps impossible, it forces one to accept that, at least in some situations, our forebears were not acting out of malice, they were acting in their interest, for better or worse, and according to the information and values of their time. People moved plants and animals for trade, and as a by-product to trade and travel.

In recognizing this, one arrives at an obvious conclusion: we have no such shield to hide behind while moving forward. Our understanding of invasive species, while not complete, is comparatively robust, and our continuing to bring about new biological invasions indicates that, rather than tackling the challenge ourselves, we would rather pass on this burden to future generations.

In writing this book we hope these themes bubble to the surface. Our key aim is to answer why biological invasions occur and much of the first two-thirds of the book is dedicated to describing the processes by which a species becomes invasive. This detail should give the reader an understanding of the technical aspect of biological invasions, but also why they occur in societal terms.

We do this by trying to show that values—that is, our relationship with the non-human world—as much as ecology underpin invasive species as a phenomenon.

All books need to be written with their audience and purpose in mind. We think we have struck the right balance in depth of topics that are central to understanding invasive species, while not getting too caught in the weeds so that the larger points are missed. Additionally, we wanted to leave enough uncertainty for the reader to be viscerally aware that invasive species are not a 'just-so' story, nor are the conclusions. Much work needs to be done on this important global challenge. A challenge of our making.

List of illustrations

Chapter 1
A global challenge

Are you carrying any fruits or vegetables? This is a familiar question encountered by the international traveller but not one you would expect to hear while driving through outback Australia. And yet, it is asked of all highway-borne traffic entering the large south-central state of South Australia. The interest in produce at these roadside checkpoints is less about the fruit and more about what could be growing in it and represents South Australia's first line of defence against invasive species.

On the western border of South Australia, the primary species of concern is *Ceratitis capitata*, the Mediterranean fruit fly, or medfly for short. It is a relatively small, squat fly, 3–5 mm in length, which creates a new generation every 3–4 weeks. The medfly is native to tropical Africa despite its name, but due to its broad environmental tolerances it has invaded numerous locations including the Mediterranean, Asia, Hawaii, Central and South America, and Australia, specifically the state of Western Australia. More than 250 plants host medfly and many are important commercial crops such as apples, oranges, grapes, and various nuts, all of which grow in South Australia. If left unchecked, the medfly could easily take up residence and devastate South Australia's horticultural industry.

Damage to a crop begins with adult female flies. They sport a sharp ovipositor, effectively a small needle extending from the fly's abdomen, which is used to pierce and lay eggs just beneath the fruit's skin. On green apples and other similarly light-coloured fruit, egg laying marks are often visible on the skin as small brownish dots. On dark fruit the marks are much more difficult to see and damage to the crop may go unnoticed until harvest. During each laying event flies deposit somewhere between six and 20 eggs (the actual number varies between individuals), which can result in more than 700 eggs being laid during a female medfly's life.

Most damage to a crop comes several days after egg laying when the eggs hatch and the larvae go to work eating the fruit's flesh. And the larvae are not alone. They are accompanied by numerous species of bacteria that help break down the flesh of the fruit. Since the larvae feed beneath the fruit's skin, even severely infected pieces may appear normal; but to the touch, the fruit will feel soft and spongy and breaking it apart will show it riddled with cavities. The larvae feed on the fruit for about two weeks before they exit the fruit, drop into the soil, and re-emerge 10 days later as adult flies to start the process anew.

Damaged portions of fruit that were grown in home gardens can often be excised, leaving the remaining piece edible, while even a mild infestation of fruit fly can be economically disastrous for commercial growers. In especially bad years farmers in Western Australia have lost half their crops to medfly, and on top of these direct losses are the associated costs of control, post-harvest treatment, and ongoing surveillance to detect fruit flies before they cause damage. Even farms without fruit fly infestation, but in areas affected by fruit fly, suffer losses as domestic and international buyers cut or cease buying from affected regions to reduce the risk of importing potentially contaminated produce. In general, pest and invasive fruit fly species cost the Australian horticultural industry approximately US$200 million annually, roughly 2–3 per cent of the annual harvest. The South Australians, with their billion-dollar horticultural industry, which includes more than 50 per cent of Australia's wine grape production, and all of which is currently fruit fly free, are well justified in their seemingly rigid policies against out-of-state produce.

Medfly of course is just one example of an invasive species, which, as a group, have become a global socio-environmental challenge. Invasive species are found in all oceans and on all continents including Antarctica. They are represented by familiar organisms such as plants, animals, and fungi, as well as microbes such as bacteria and protists. And their impacts vary from economic losses, which ecologist Christophe Diagne and colleagues recently estimated to be US$162.7 billion dollars worldwide in 2017, to losses of biodiversity and ecosystem services. Invasive predators alone have contributed to 58 per cent of bird, mammal, and reptile extinctions globally, while just a single invasive tree, *Melaleuca quinquenervia*, has dramatically affected the hydrology and fire regime of southern Florida.

Some argue that these impacts are natural processes. Organisms have colonized novel habitats while others have gone extinct since life began. And this is true. The concern with invasive species is not one of novelty, but primarily one of spatial extent and magnitude. The global trade network coupled with modern transportation makes any specific location biologically connected to historically separate regions. Fruit that is picked, packed, and posted in Hong Kong today could be in London, New York, or Sydney within 24 hours. Along with this food product comes any species that are present in or on the fruit, like fruit flies. The rates of species' colonizations on islands, which were historically 0.2 or fewer species per year, now range between 20 and 35,000 species per year. Aquatic environments have seen similar increases in colonization rates. The Caspian Sea, as large or larger than many European countries, experienced an 1,800-fold increase in the number of newly established aquatic species over the 20th century due to human activities. Ecosystems simply cannot adapt quickly enough to this rapid influx of new biota.

This modern-day pattern of greatly accelerated colonization has real and often major consequences for human livelihoods and the ecosystems into which these species establish. It is not too strong to say that invasive species, while fascinating for what they reveal about ecosystems, are one of modern societies' greatest challenges and greatest failures. To meet this challenge, we must understand how organisms become invasive, and the first step in reaching that understanding is to answer the question: w*hat is an invasive species?*

Chapter 2
Nature abhors a definition

Nature resists being squeezed into a box and definitions of biological entities typically contain many caveats. For example, more than a dozen concepts attempt to explain what is and is not a *species* and determining what constitutes life itself has led some to conclude that such a classification is pointless. Although these definitions do not unambiguously classify all biological entities, they do provide practical utility for steering research, evaluating new findings, and developing management responses. The same applies to the definition of what constitutes an invasive species. It is a tool that works most of the time for most organisms in most places, but there will always be borderline cases that undermine it and incite debate among scientists, policy makers, and the interested public.

Before we define invasive species, we should note that other terms are often used in the public, political, and even scholarly literature to refer to 'invasive species'. Alien species, exotic species, feral species, or even weedy species might be used as synonyms for 'invasive species'. Part of this variation is historical with some terms originating in particular fields of study; 'weedy species' is a term often seen in the botanical literature, for instance. Other terms, such as 'feral species', technically refer to a subset of 'invasive species'. As you will see, however, how we define invasive species is much more about us, how we interact with nature, and our values, than it is about the organism being labelled *invasive*. With that said, we can set out a working definition: an invasive species is a population of non-native species that either cause undesirable impacts or that have spread beyond their initial introduction location. We will now briefly unpack the major terms to alleviate any ambiguity.

What is a biological population?

A *population* refers to those individuals of a species that inhabit the same geographic region and it may constitute all individuals of that species, but often does not. For example, Burmese pythons (*Python bivittatus*) in southern Florida constitute a different and invasive population from Burmese pythons found in the species' native range of south-eastern Asia even though they are clearly the same species. If we are talking about *populations* of a species and not the species in its entirety, then why call them invasive *species*? Because saying 'an invasive population of a species' is rather long-winded, and typically when we refer to an *invasive species*, it is often in the context of a specific case; for example, 'the Burmese python is an invasive species in southern Florida'. Throughout this book, when we mention *invasive species*, keep in mind that we are talking about a population and not all individuals of the species.

What are non-native species?

All biological populations can be broadly classified as either native or non-native. The distinction between them is principally determined by our role in an organism's distribution. If humans help a species across a biogeographic barrier to a location outside its native range, regardless of whether our action was deliberate or accidental, then that species is not native in that new region. There are, of course, caveats to this.

Biogeographic barriers are naturally occurring geographical or ecological features that individuals of a species typically cannot cross, which restricts a species from exploiting all potentially suitable habitat. Mountain ranges, oceans, canyons, and other physical features are common examples; and what is insurmountable, of course, greatly varies between species.

Since biogeographic barriers rise and fall on time scales of decades to millennia, instances of individuals of a species crossing such barriers have been comparatively uncommon, at least until human intervention. Many modern non-native species have crossed barriers because we moved them, either knowingly or accidentally, while other species crossed their barriers when we substantially modified landscapes. In the most extreme cases, we remove the barrier almost entirely. After the construction of the Suez Canal in the 19th century, for instance, marine species from the Mediterranean Sea could reach the Red Sea, and vice versa, and some of these dispersing species have become non-natives in the others' home.

There are two caveats to a species being considered 'non-native' despite their dispersal being facilitated by people. The first hangs on when the facilitation occurred. Ever since there were humans, we have aided in the dispersal of non-human organisms. For much of our history, however, our contribution was akin to that of other non-humans. A bird might consume a plant's fruit on one island, fly to another, and deposit the seeds during defecation; and humans similarly moved plant species across similar barriers. By modern standards, these events were comparatively uncommon, involved few species, and often occurred over short distances.

Our contribution changed dramatically in the 15th century (Table 1). Starting with European-led colonization activities, ships filled with various plants and animals struck out for new horizons and returned with foreign counterparts. Numerous species successfully established non-native populations in new lands during this great biotic exchange. The best known perhaps is the ship or black rat (*Rattus rattus*). Its proclivity to stow away on

Table 1. Differences between ancient and modern facilitation of non-native species to new regions by humans

Characteristic	Ancient facilitation	Modern facilitation
Frequency of long-distance dispersal events	Very low	Very high
Number of species transported per event	Often few	High
Number of organisms transported per event	Often few	Very high
Effect of biogeographic barriers	Strong	Weak to insignificant
Variation in mechanisms and routes of dispersal	Low	High
Temporal and spatial scales of mass invasion events	Episodic, limited to adjacent regions	Continuous, affects all regions simultaneously
Homogenization effect	Regional	Global

ships and in cargo of all kinds has extended its distribution from Europe and Asia to all continents except Antarctica. Trade networks and technology have only improved since the 15th century, and with them so has the frequency of non-native species' introductions. Therefore, a working rule-of-thumb has emerged whereby the year 1492 is considered a reasonable date to distinguish between ancient versus modern facilitation events and, by extension, what is considered native or non-native.

A second caveat requires that the suspected non-native species exists as a self-sustaining population. That is, the non-native population must be free-living and reproduce without human input. This definition excludes species being kept in captivity, such as domesticated or cultivated species, from being described as 'non-native'. Nevertheless, if a tribe of domestic goats (*Capra aegagrus hircus*) escapes a farm and establishes a wild population, then they would qualify as a non-native species; and if said population then caused damage to crops and so forth, they would be considered invasive. This last caveat also excludes labelling an individual, or even several individuals of a species, outside their native range as non-native if they are unable to generate a viable population. Such individuals are typically termed 'introduced species'.

What are undesirable impacts?

By virtue of interacting with the surroundings, all organisms have an impact on their environment. Some of those impacts, whether the species is native or not, may be considered 'undesirable'. Native termite species for instance are, globally, one of the most damaging pest species (native species that cause undesirable impacts); can their impact on houses and public infrastructure be considered as anything but 'undesirable'? The undesirability of an impact depends upon the value that we, as a society, place on those things being impacted. And this is the key point; the concern with invasive species is not that their impacts have a 'negative effect' on the environment or another species' population, as in they reduce that other species' population—the issue is that invasive species affect the value society gains from the thing being impacted. When fruit flies reduce a farmer's harvest, we determine the negative impact on crop yield to be undesirable because we value crop yield. Similarly, the Burmese python invasion in southern Florida has caused a decline in native species' populations, which is undesirable because we value larger populations of native species.

Because 'undesirable impacts' are predicated on values, what one group of stakeholders might find undesirable, another group may not. In a few cases, one person's invasive species may be another's desired resource. This is a point that we will return to later in the book—still, we should not interpret 'undesirable impacts' nor, by definition, what is considered an invasive species as being arbitrary. Society's values towards the human and non-human environment certainly change through time, but they do so largely in accordance with our understanding of those domains. As our understanding of the importance of healthy ecosystems for human wellbeing has grown, so has the value we place on maintaining healthy ecosystems.

Why spread is sufficient

In the context of invasive species, spread refers to a non-native species dispersing from their initial introduction location and establishing a more widespread distribution. Including spread in the definition of 'invasive' provides an ecological grounding to the designation, which is useful for structuring research on why and when some non-native populations expand their distribution where some do not. Ordinarily, undesirable impacts and spread go hand in glove; the more widespread a species, the more likely it will cause undesirable impacts (we explore this in detail in Chapters 6 and 7). Still, the condition of spread is included in the definition of invasive species even if a widely distributed non-native species may not, at the time it is noticed, cause any obvious undesirable impacts. To this, one might ask, if the organism is not causing harm why should we consider it invasive? There are two important reasons for this.

First, one particularly puzzling aspect of biological invasions is the concept of 'invasion-lags'. This term captures several phenomena observed in the invasion process, one of which refers to impact-lags, the process where a non-native species is present for some time before impacts occur at a level sufficient to be observed or measured. It follows then that a non-native species not currently causing undesirable impacts is not a good predictor of future undesirable impacts by that species. And, second, recall that the purpose of an invasive species' definition is to steer research, management, and policy. Identifying a widespread non-native species as invasive, even in the absence of it causing overt undesirable impacts, ensures that it remains under scrutiny by the research and management community. Waiting for a species to cause undesirable impacts before taking further action is not a good environmental management solution, especially when some impacts are potentially irreversible.

The concept of an 'invasive species' is only about 60 years old and began with British zoologist Charles Elton's publication *The Ecology of Invasions by Animals and Plants*. Elton's work grew out of concerns being raised in the late 19th century that the widespread introduction of all manner of species to novel habitats may be environmentally disastrous. In innumerable cases, those concerns were warranted. Given this relatively recent concern about invasive species, our understanding of them is both relatively new and, like our understanding of ecological systems generally, far from fully resolved. Still, many of the fundamental concepts are in place and the coming chapters explore how a species becomes invasive.

Chapter 3
Pathways of introduction

The invasion process is characterized by three main processes: introduction, establishment, and spread. That is, for a species to become invasive it must be introduced to a location in which it is non-native, establish a self-sustaining population, and if it hasn't already caused undesirable impacts, spread (Figure 1). The invasion process can be viewed as a series of filters. Not all species that could be potentially introduced to other environments are introduced; not all introduced species go on to establish

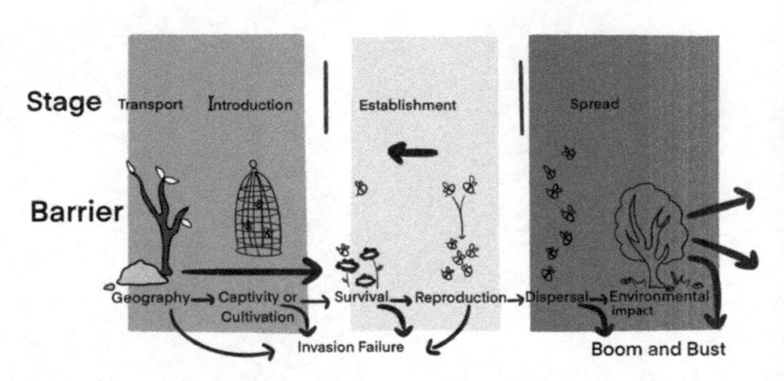

1. The three main stages to an organism becoming invasive: transport and introduction, establishment, and spread. At each step, the non-native species has to overcome barriers such as surviving the transport process and making it into the wild, surviving, and finding a mate to establish a population.

non-native populations; and not all established non-native populations will spread or cause undesirable impacts. We will examine establishment and spread in the following chapters. Here, we focus on introduction, which captures the transport of a species from one location to another and its subsequent entry into that environment.

All introductions are a result of human activity, and primarily due to our trade and transport networks. While some species are introduced to novel environments because we deliberately release them, other introductions are accidents, the unintended consequences of trade when moving people or products around. The number and types of species introduced throughout history has waxed and waned as our knowledge of ecological systems has increased, as industries have changed, and as technology has evolved. For example, the number of non-native mammal introductions per year has decreased since the 1950s, when their undesirable impacts were becoming evident, and yet since then we have seen an exponential increase in insects being accidentally introduced due to the expansion of global trade (Figure 2). So, in the year 2000, more than 2,000 invertebrate species were introduced to new regions globally, and a high proportion of those established a population. The graphs in Figure 2 in fact understate the total number of introduction events since repeat introductions of a non-native species to a region are not included. The same species might be introduced several times at a location, and only be counted as a single entry in this figure.

The variability in the types of species introduced, as well as the number of individuals of a species introduced, are best explained through five introduction pathways: corridor; stowaway; contaminant; escape; and release. These pathways grade according to how direct our actions were in facilitating the introduction. On one end, the 'corridor' pathway accounts for species that disperse to new ecosystems on their own accord after we have modified the environment, while on the other end, the

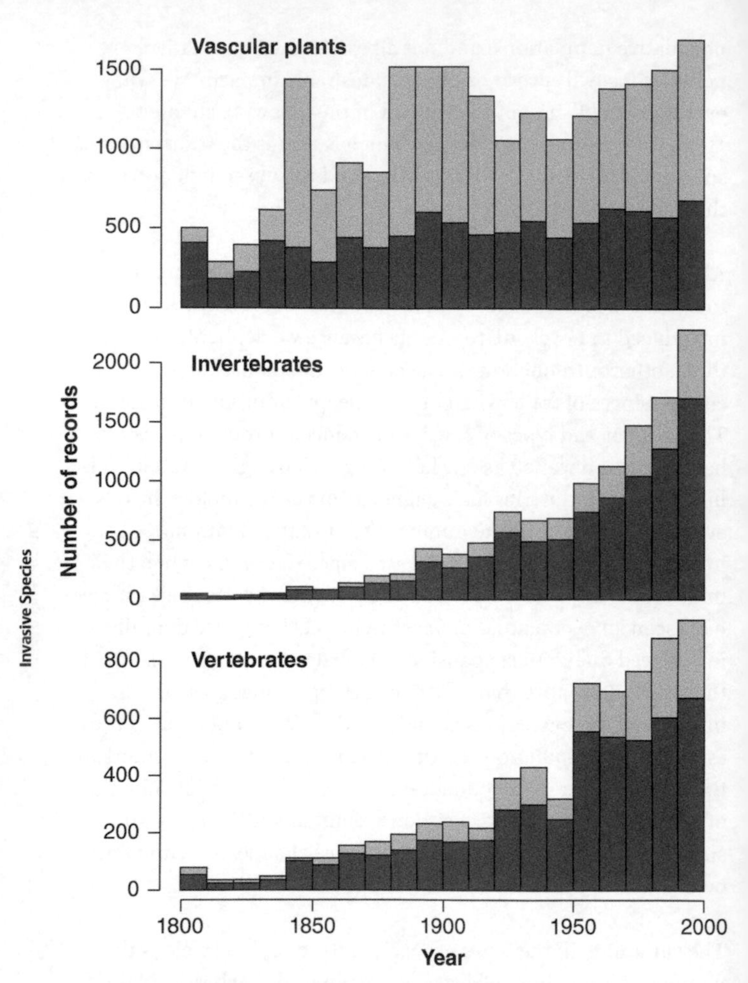

2. The global number of recorded non-native invertebrate and vertebrate species that were first introduced (light bars) and those that are recorded as becoming established or invasive (dark bars) has grown exponentially since 1800, while new introductions of vascular plants have remained somewhat constant since 1900.

'release' pathway accounts for those introductions where we deliberately released individuals.

Where there's a corridor there's a way

Movement corridors are pathways that organisms can use to transit between patches of suitable habitat. Although some terrestrial species are introduced via corridors, most species in this pathway are aquatic and exploit canals such as the Suez or the Panama to disperse between parts of the ocean. Constructing a canal removes a major biogeographic barrier for marine species, but it does not necessarily remove all barriers. The Suez Canal traces a path through the highly saline Bitter Lakes, which no doubt impedes many organisms from journeying between the Mediterranean and Red Seas. Still, crossing a salinity barrier is far easier than crossing a land barrier for marine species. The former requires a tolerance to high salinity whereas the latter requires the ability to move, breathe, and survive on land. As such, approximately 400 marine species, including more than 100 fishes, native to the Red Sea are now established in the eastern Mediterranean.

Compared with the Suez, the Panama Canal is a much more formidable barrier for marine species, but it is not impermeable. A series of locks raise the main water level up 28 metres from the Pacific Ocean to the fresh waters of Lake Gatun, then back down again to the salty Caribbean Sea. The locks themselves pose little obstruction to many marine organisms, and euryhaline species, organisms that can tolerate a range of salinities, have transited the canal or taken up residence along the way. The best-known traveller is the Atlantic tarpon (*Megalops atlanticus*), which was first documented as having a non-native population on the Pacific coast of Central America in the late 1930s, only 25 years after the canal's construction. Due to the mixing of fresh and salt water during canal operations, Lake Gatun has become increasingly more saline and approximately 30 marine fishes have established non-native populations there.

We generally never intend to create introduction corridors when we construct shipping or transportation corridors. But the removal of major biogeographic barriers will allow some species to disperse and establish non-native populations.

Hitch-hiking to invasion

There is an easier way to traverse biogeographic barriers than climb them yourself: hitch a ride. Species using the stowaway and contaminant pathways do just that on human transport. In both pathways, species are transported accidentally, and reach new environments without us knowing. The difference, however, is that contaminants are those species associated with specific products that we are transporting, such as fruit fly being introduced with imported fruit or parasites and diseases that accompany pets. Stowaways, on the other hand, refer generally to any non-native species accidentally entrained within human transport. An infamous example of the latter is the brown tree snake (*Boiga irregularis*). It was accidentally introduced to Guam in the early 1950s after being transported with military supplies out of its native range in the south Pacific.

Species in the contaminant and stowaway pathways are primarily small organisms such as invertebrates, plant seeds, fungi, and microbes. The specific species and the distances they are moved have changed as transport practices and technology evolved over preceding centuries. One of the greatest changes in the range of species transported occurred when ships shifted from dry to wet ballast at the start of the 20th century. Ballast has long been used to improve stability and efficiency of vessels and before the 20th century soil, rock, and other dry ballast materials were hand loaded into a ship's holds along with cargo. As ships sailed from port to port, ballast requirements changed. Unnecessary ballast was dumped at the docks and ballast needs were met by those dumps, or from whatever materials were at hand. The movement of dry ballast was the cause of hundreds, even thousands, of plant and invertebrate species being introduced to terrestrial environments around the world.

Around the 1900s, vessels began to use wet in place of dry ballast by pumping the surrounding water into tanks within the ship. This method is much more efficient than loading dry ballast by hand as tanks can be filled and emptied without needing to be docked. Filters stop large organisms from being ingurgitated into the tanks, but do not filter out the small larvae of large organisms, nor zoo- and phytoplankton, bacteria, or viruses. These organisms are subsequently sucked into the ship's tanks in one location and, as with dry ballast, discharged at another location when ballast requirements change. So, although ships stopped dispersing plant seeds and terrestrial invertebrates in dry ballast, they began to disperse algae and all manner of aquatic invertebrates in wet ballast.

Ballast-mediated invasions have devastated some ecosystems. The warty comb jelly (*Mnemiopsis leidyi*), for instance, a 10-centimetre-long ctenophore native to the western Atlantic, was introduced into the Black Sea in the early 1980s from ballast water. Without any predators its population rapidly increased and reached densities as high as 300 individuals per cubic metre in the Black Sea. Since the comb jelly feeds on zooplankton, a food source shared by anchovies (*Engraulis encrasicholus*), the massive comb jelly population dramatically reduced zooplankton densities and ultimately contributed to the collapse of the anchovy fishery in the Black Sea region.

Since the late 19th century, we have seen an alarming increase in introductions via the contaminant and stowaway pathways due to two key developments in transportation. First, trips between locations are now made with more frequency and pace than ever before. The pace at which people, goods, and hitch-hiking non-native species move, therefore, is now measured in hours and days rather than weeks or months. Shorter duration trips increase the probability that non-native stowaways survive being transported, while an increase in trip frequency sees more non-native individuals being entrained within the invasion process. And, second, our air, sea, road, and rail networks have become more interconnected and together now exhibit a classic 'small world' characteristic, whereby any location can be reached from any other in a few steps. Our transport networks have effectively become superhighways for non-native species as almost all regions are now biologically connected to almost all others.

Not all 'roads' on these superhighways are equivalent, however. First, although some locations are connected to a few others, some are connected to many others due to the classic 'hub-and-spoke' model adopted in transportation networks. These hubs effectively form 'invasion bridgeheads', places that are the source of many non-native species introductions, often because they host many non-native species themselves. Around 80 per cent of non-native ants detected entering the United States and New Zealand between 1914 and 1984, for instance, have come from areas (hubs) outside the invading ant's native range, areas that the ant had already invaded. Invasions beget invasions as it were.

Another inequality is that connections are largely unidirectional. Despite heavy bilateral trade between Pacific and Asian nations, the 'flow' of non-native species is mostly from Asia to New Zealand and Australia, and not vice versa. In the extreme, highly connected islands such as Hawaii receive non-native species from nearly every corner of the globe but seldom are Hawaii's native species introduced elsewhere, perhaps because the evolutionary adaptations to life on an island make these species poorly adapted to life elsewhere.

Not all organisms travelling via human-made corridors survive their journey, yet the ones that do exhibit characteristics that aid their survival and often make them better invaders in new ecosystems. This pattern is the same for species in the stowaway and contaminant pathways. Conditions en route impose some level of stress, dehydration, desiccation, or starvation on the organisms. Marine species sucked into ballast tanks, for instance, are plunged into darkness, experience temperatures that diverge from their norm, are exposed to higher concentrations of competitors and predators as well as chemical contaminants such as oil, and the journey may last weeks. The individuals and species that survive such journeys therefore are those with physiological adaptations to survive heat stress and periods of food scarcity, and most likely are more competitive in their new environments. The conditions in the holds of planes and in trucks are less hostile, but still some number of individuals will perish, only to be survived by the more robust. We know this because in the next two pathways, where we deliberately transport organisms, not all survive their journey either.

Putting the best species in the best places

We have now arrived at the other end of the introduction spectrum, where non-native species in the final two pathways of 'escape' and 'release' reach new regions because we intended to move them. The specific reasons for doing so are nearly as varied as the species themselves but, generally, the underlying motivation has always been financial gain. Many species are transported with the intention of maintaining them in captivity, such as reptiles for the exotic pet trade, but for various reasons such organisms escape their confines or become unwanted and are released by their owners. Other species were transported with the explicit intent of releasing them to the wild to control other invasive species, or for recreational fishing and hunting, or attempts to start new industries. Where the stowaway and contaminant pathways are mostly populated by small species, often invertebrates, most vertebrate introductions were a result of the escape and release pathways. And here we have at least learnt a lesson of a kind.

From the 15th through to the start of the 20th century, during the era of European expansion and colonization, moving flora and fauna to where we saw fit was the norm. The first few centuries, often referred to as the Columbian Exchange, marked the beginning of an immense movement of organisms between the New and Old Worlds. Much of the exchange comprised agricultural species such as corn, tomatoes, potatoes, and various species of cattle. During this period, explorers would also transport and introduce domesticated species such as goats, pigs, and sheep to new regions to establish sources of food for future voyages. The British explorer Captain James Cook is credited with introducing goats to islands around Australia, New Zealand, and Hawaii during his 18th-century voyages, and a number subsequently established free-living populations, resulting in the loss of some plant species in these locations.

The deliberate introduction of non-native species, especially those without an agricultural purpose, peaked in the mid-19th century with Acclimatization Societies. The societies, present in many countries, endeavoured to introduce any species to any area if it was thought to be beneficial, with 'benefit' narrowly defined by the society's members, who included politicians, landholders, and naturalists. By today's standards, reading the transcripts of society meetings provokes laughter and dismay. In Australia, at the Acclimatization Society of Victoria's annual general meeting in 1863, the then Governor of Victoria, Sir Charles Darling, suggested that boa constrictors should be introduced to the colonies because they would supposedly eliminate venomous snakes. It's not clear how this would have worked since the primary diet of boa constrictors is mammals. Although this suggestion fell flat, many others were heeded with disastrous effects.

We should not be too hasty to condemn 19th-century societies for their ecological missteps. They were working with the knowledge of their day and 150 years of continued blunders and scientific development has provided us with an ecological understanding unavailable to them. By the end of the 19th century the lustre of numerous plant and animal introductions was tarnished as promises of benefit never bloomed, or worse, the introductions became irritants. In his autobiography *My Life in Two Hemispheres,* Sir Charles Gavan Duffy recounts a fellow parliamentarian in the Victorian state government (Australia) in the early 1870s lamenting:

> Another member whispered, 'Let us alone with your new industries. You see what has come of them already. A Scot introduced their charming thistle, and we will have to put a sum on the estimates to extirpate it. Edward Wilson introduced the sparrow, and the sparrow is playing havoc with our vineyards. Some busybody introduced the rabbit, and the income of Ballarat would not save us from the consequences.'

Some in the United States also looked to slow introductions. In the mid-1890s, the American zoologist and then Assistant Chief of the United States Department of Agriculture, Theodore Sherman Palmer, authored *The Danger of Introducing Noxious Animals and Birds* and concluded, 'The introduction of exotic birds and mammals should be restricted by law.'

While the deliberate introduction of flora and fauna, especially birds and mammals, has decreased since the 19th century, lessons have been learnt the hard way. Throughout the 20th century, and even today, many countries have continued to release non-native species in pursuit of economic gain. Nutria (*Mycastor coypus*), a large, herbivorous, semi-aquatic rodent, was introduced to North America from its native range in South America for the fur trade in the 1930s. In the 1940s the North American beaver (*Castor canadensis*), another large, herbivorous, semi-aquatic rodent, was introduced into South America from its native range in North America, also for the fur trade. And now, in both cases, expensive attempts are under way to undo those invasions. Modern deliberate introductions are most noticeable in the service of hunting and fishing industries where species, such as rainbow trout (*Oncorhynchus mykiss*), are bred and released annually in several countries, where the species is clearly not native, to sustain the fishing industry. Fortunately, insects are rarely introduced deliberately, and when they are it is for biocontrol—species used to control other species—and only after extensive experimentation to ensure they do not run amok.

Vertebrate organisms and ornamental plants are now mostly introduced through the 'escape' pathway in which species are deliberately transported with the intention of maintaining them in captivity. These organisms are typically commodities for the ornamental plant and exotic pet and aquarium industries and become free-living by escaping their confines, or when their owners dispose of them inappropriately. Both the Burmese python (*Python bivittatus*) and the Pacific lionfish (*Pterois volitans*), for instance, were imported to Florida in the United States to feed consumer demand for exotic pets. And in both cases, they entered the wild and subsequently established invasive populations when people released them. While this is a form of 'release', the intention in releasing the organism most likely was not to introduce it, as is the case in the 'release' pathway; instead the organism was released because the owner concluded that they had no other suitable means of disposing of the animal.

An individual pet escaping or being released into the wild has little impact on an ecosystem's long-term health if it quickly dies or fails to find a mate. But pet and aquaria species do not exist in isolation or in only a few households. Over half of all households in the United States, the United Kingdom, and Australia have pets, and roughly 50 per cent of these are not domesticated species such as dogs, cats, or horses. Somewhere between two and five million individual birds, representing approximately 2,500 species, are sold globally each year. Fish are among the most traded pet species. More than 11 million individual marine fishes from 2,300 species are imported into the United States annually, and freshwater fish are an order of magnitude greater again. Most individuals will remain in captivity but, even if only one in a hundred make it to the wild, the pet trade represents the source of hundreds of thousands of non-native organisms entering ecosystems globally each year.

Figure 3 is adapted from research done by John Wilson and colleagues, who examined the importance of various introduction pathways for non-native species being introduced to South Africa since 1700. The first striking feature, even for this one location, is that no group of organisms have avoided being entrained in the transport and introduction phases of invasion. And, second, we get a clear sense of how introduction pathways have ebbed and flowed through time for different groups of species, the shift from dry to water ballast in the 1900s that has greatly facilitated marine introductions, and the increase in reptile introductions due to the increasing influence of private keeping practices. Perhaps the most striking and disappointing feature is that the estimated importance of most introduction pathways post 2000 is not rapidly declining. This will result in the continued introduction of non-native species to new locations in the future.

Not every organism that starts down an introduction pathway survives of course. Many perish due to the conditions en route. This outcome is especially common for those organisms that transit via corridors or within human goods since conditions in those pathways are often far outside these organisms' nominal environment. But the variability of the pathways coupled with the adaptability of life, and simply the sheer number of individual organisms entering the various pathways, means that many individuals of many species do survive and reach new ecosystems. And it is from these survivors that new populations have a chance to establish.

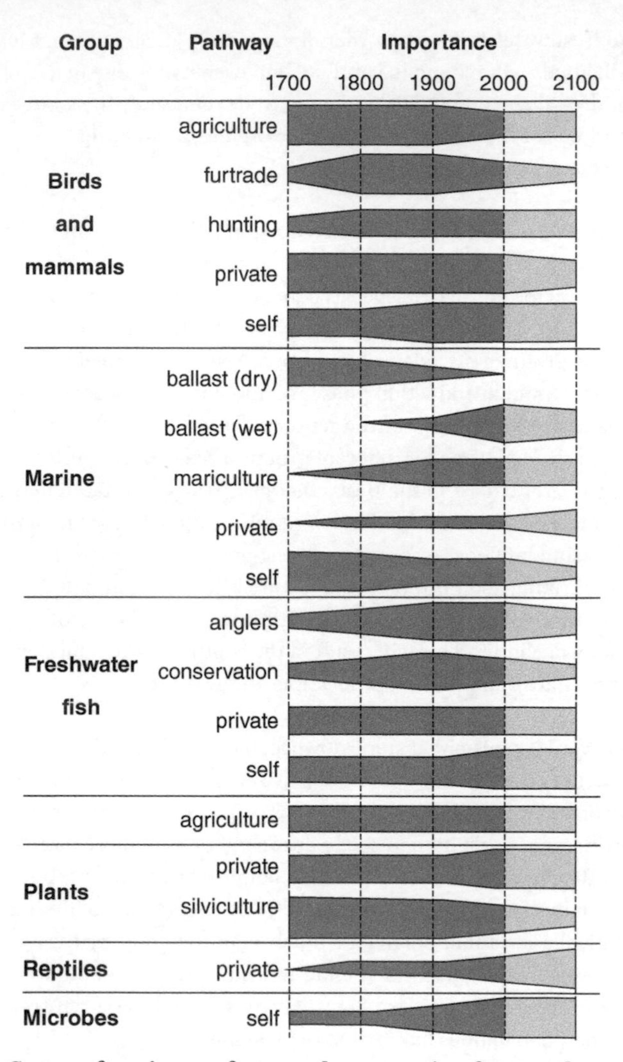

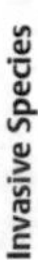

3. Groups of species are often moved as non-natives for more than one reason (pathway). The importance of these pathways in terms of the number of non-native species moved has varied over the last three centuries, here depicted as the width of the lines associated with each pathway.

Chapter 4
Establishing a population

Having successfully traversed an introduction pathway and made it to a novel ecosystem, the prospective invasive species now must establish a population. At this point, it is merely an 'introduced species'. It now needs to survive, thrive, and reproduce if it is to establish. And, if it does successfully reproduce, then the following generation too needs to survive and reproduce, and so on and only then will the introduced species have established a persistent population.

Establishment is far from a foregone conclusion for the newly introduced then. Surviving the introduction pathway does not guarantee survival in the new habitat, as the latter depends on whether the species' niche is available in the new habitat. It may not be, and the potential invader would almost certainly perish as a result. The difficulties are compounded for sexually reproducing species as they need to find a suitable mate to establish a population, and there may be only a few individuals introduced into any one location at any one time. Fortunately for them, we often help by continually introducing new individuals into the same locations year after year.

N-dimensional hypervolume

All organisms have a niche. It is the set of abiotic and biotic resources that are required for that organism's survival. To make this concept more concrete, consider a hypothetical plant that lives in a subtropical environment and can tolerate freezing temperatures. Temperature is one dimension of that plant's niche and we can plot it against suitability for the plant on the y-axis of a graph (Figure 4). Within a certain range of temperatures, in this case 10 to 30 degrees Celsius, most individuals of this plant will survive, thrive, and reproduce. Outside this range, life becomes progressively more difficult, until temperatures are so hostile that no individual of the species can survive. This is a graphical depiction of our plant's thermal niche. Each species will be different; some with the same optimal temperature, but with broader or narrower tolerances. Some might be more suited to higher temperatures, while others thrive in lower temperatures. And temperature is only one dimension. We can continue adding dimensions, represented by axes on our graph, to account for precipitation, solar radiation, nitrogen, salinity, and so on until we reach what the British ecologist George Evelyn Hutchinson termed an '*n*-dimensional hypervolume': the set of environmental characteristics that an organism requires to survive.

Introduced species often fail to establish due to a mismatch between their niche and the environment into which they were introduced. This dependence is particularly evident in the aquarium industry. Hundreds of millions of fish, representing thousands of species, are sold as pets every year, and somewhere between 3 and 10 per cent of people who purchase them will eventually release them into nearby waterways when fish care becomes burdensome. One study estimated that around 10,000 fishes are released annually into the waterways around Montreal, Canada. Just five species accounted for the vast majority of these released former pets, including the ever-popular goldfish

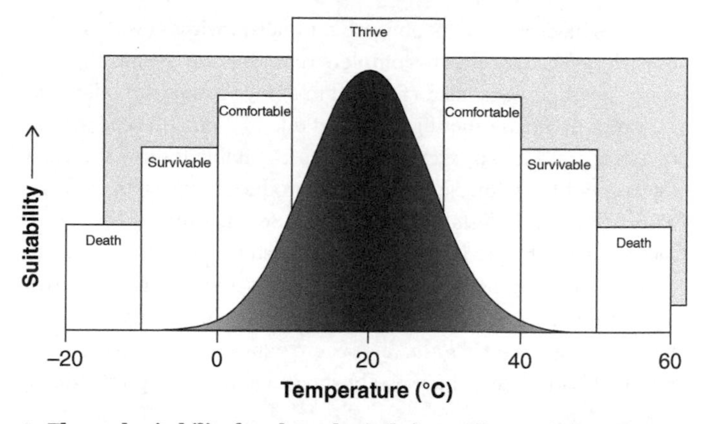

4. Thermal suitability for a hypothetical plant. All organisms occupy a niche, of which temperature is one dimension. Here, the hypothetical plant cannot survive in temperatures below -10 and above 50 degrees Celsius, with temperatures between these extremes becoming more suitable to an optimum temperature of around 10–30 degrees Celsius.

(*Carassius auratus*) and neon tetra (*Paracheirodon innesi*). Despite the large numbers of these species entering these waterways, none of them has established near Montreal. This failure is because all are native to relatively warm waters—goldfish are from east Asia and the tetra originates in the western and northern Amazon—whereas most rivers and lakes around Montreal drop to near or below freezing each winter. These species simply cannot tolerate such low temperatures and released individuals likely perish each winter.

An organism's niche also extends to biotic dimensions such as food resources or, where it is seen most clearly, when symbiotic relationships are required for reproduction. Consider the Chinese banyan fig tree (*Ficus microcarpa*). It is a spectacular tree with large, sprawling roots that mirror their crown and, for that reason, has been distributed nearly worldwide for ornamental gardens. Yet, the Chinese banyan has not established in all places in which we have planted it. The issue here is not climate, but pollination.

Fig trees have evolved an obligate mutualism with fig wasps; that is, each needs the other to complete reproduction. Female fig wasps lay their eggs inside figs and in doing so deposit pollen that they pick up from other fig trees, but due to their co-evolved histories, not just any fig wasp species is suitable for any species of fig tree. For the Chinese banyan to reproduce it needs its fig wasp, *Eupristina verticillata*, for pollination. So, the Chinese banyan has rarely established where it was planted because its wasp was absent. This circumstance has changed in several places recently as continued importation of Chinese banyans has brought hitch-hiking *E. verticillata*, and where banyan and wasp are reunited both species have established as non-native populations.

Biotic resistance

Even when a habitat exhibits suitable conditions, introduced species may not establish due to 'biotic resistance': native species outcompete the invader for limited resources or native predators preferentially eat the invader. One of surprisingly few examples of biotic resistance is seen in the introduction of the green crab (*Carcinus maenas*) to the United States. The green crab is native to the European and North African Atlantic coasts and they have been transported and introduced the world over for several hundred years—hitch-hiking in dry and wet ballast and on the bottoms of ships—and have consequently become established in several countries, including estuaries along the Atlantic and Pacific coasts of the United States. However, green crab abundance in both east and west coast estuaries varies considerably because, in part, they are eaten at differing intensities by larger native crabs. On the east coast, green crabs are found at significantly lower abundances where native blue crabs (*Callinectes sapidus*) are found, compared with where blue crabs are absent. A similar story plays out on the west coast with the native red rock crab (*Cancer productus*), although on the west coast green crabs are more abundant in part because red rock crabs are less likely to eat them.

Traits to survive establishment

The corollary to these niche and biotic barriers is that many species that do establish often exhibit one of the following traits. First, established species often exhibit a broad niche, that is, they can survive a wide variety of ecological conditions. The common starling (*Sturnus vulgaris*), for instance, exploits tropical, temperate, and boreal habitats. Due to its numerous and widespread introductions combined with its broad niche, the starling has established populations in North, Central, and South America, Australia, New Zealand, several Pacific islands, and South Africa in addition to its native range of Europe and west Asia. In the world of invasions, generalists like starlings are common.

Second, numerous introduced species establish because they exploit habitats that native species do not. This situation is commonly observed where humans have altered the environment beyond levels native species typically tolerate. For example, modern agriculture employs large quantities of nitrogen and phosphorus, nutrients foundational to nearly all organisms on Earth, but which are naturally scarce. Most species have adapted to nutrient scarcity and do not readily tolerate the nutrient-rich terrestrial and aquatic environments within agricultural regions. Introduced species that can take advantage of nutrient-rich environments, such as black mustard (*Brassica nigra*), river tamarind (*Leucaena leucocephala*), and Asiatic honeysuckle (*Lonicera japonica*), which were imported for ornamental gardens, can and do exploit these 'unused' resources.

A third trait common to many established species is captured in the concept of 'enemy release'; the condition in which the introduced species finds themselves without 'enemies' such as predators, competitors, or parasites in the non-native habitat. An introduced species without enemies, or at least with many fewer, has a big advantage. Its daily efforts are now purely focused on putting down roots, and not spending precious resources defending itself or avoiding enemies. Sticking with plants, one study found that of 473 plant species that have established outside their native range, 77 per cent of them in the non-native range had fewer pathogens and fungi than their native range brethren. Escape is, nevertheless, often only temporary for these non-native populations. Native predators, competitors, and parasites often adapt to the invader and the number of enemies that the invader encounters increases the longer it has been established. Non-native species, therefore, can escape their enemies for a while, and perhaps just long enough to help ensure successful establishment, but not indefinitely.

Propagule pressure

Even if the invader's niche is perfectly accommodated within the new environment, and few species that serve as competitors, predators, or pathogens to them exist in this location, the introduced species faces a final hurdle before it can successfully establish: reproduction. A single introduction event of a male and female, or a single gravid female or individual from an asexual species, could lead to the establishment of a non-native population. But the final commonality between most established species is that establishment results from numerous individuals being introduced, often over multiple introduction events. These measures, together referred to as 'propagule pressure', are among the most consistent predictors of long-term establishment success. Put simply, high propagule pressure equates to a high chance of establishment success.

The reasoning is quite intuitive. As the number of founding individuals of an introduced species increases, so too does the probability that viable mates will live long enough to find one another, reproduce, and thus establish a population. However, if there are few founding individuals, the establishing population will be highly vulnerable to localized disturbances. A fire or flood could easily wipe out a small nascent population, for instance, or decrease the number of individuals, making finding a mate and reproduction rare. With increased propagule pressure, that is, with a greater number of founding individuals, the nascent population becomes more resilient, bolstered against external shocks, and more likely to establish a persistent population.

How many founding individuals are enough then? The propagule pressure required for successful establishment varies between different species and even for the same species in different locations, but by tracking the history of establishment successes, Phillip Cassey and colleagues determined that around 100 founding individuals provides an 80 per cent chance of

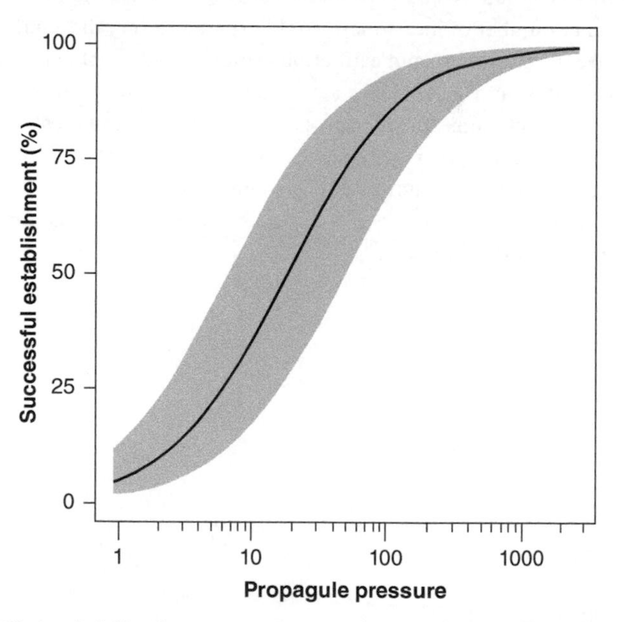

5. The probability that a nascent non-native population will establish a persistent population increases as the number of founding individuals (propagule pressure) increases. The shaded area around the dark line represents the 95 per cent confidence interval of the estimate.

establishment (Figure 5). Establishment success increased most rapidly when propagule pressure increased from 10 to 100 individuals, after which additional founding individuals in one location had little effect. With modern transport and trade networks, the introduction of 100 individuals to a single location is not generally problematic.

Introduction pathways lead to a complex geography of introduction events with individuals from the same species being introduced to multiple locations. The environmental and ecological characteristics of those habitats will almost certainly vary, and one location may be more suitable than another. It follows then that two nascent non-native populations, founded by a similar number of individuals, could experience very different chances of extinction, and extinction's complement, establishment success. It is quite possible for a non-native species with several nascent populations spread across a wide region to have most of these populations go extinct, leaving only one to persist. This one persistent population, however, is enough to establish the non-native species in the region and provide the opportunity for further spread and impact.

Chapter 5
How populations spread

Macropods are a curious family of marsupials, the most familiar of which are kangaroos and wallabies. They have shunned running on all fours for jumping when they want to cover distance. There are few places in the world to see macropods in the wild. Of course, one place is where macropods are endemic, Australia. Another is Rambouillet forest, just over the horizon from Paris, France. There, about 100 individuals of Bennett's wallaby (*Macropus rufogriseus rufogriseus*) have established a new home, and yet, despite having established a population thousands of kilometres from where they are naturally found, they are not an invasive species there.

Bennett's wallaby was brought to France for zoological gardens, and the population in Rambouillet forest became established in the late 1970s, when approximately two dozen individuals escaped Sauvage Zoo. Fifty years on, however, sightings are relatively rare and the population has not spread far from the initial location of introduction. To become an invasive species the established population needs to cause undesirable impacts and, while this could occur at the site of establishment, such impacts are often a result of the population growing and expanding into other areas. Bennett's wallaby is certainly capable of becoming invasive. Three to five pairs of Bennett's wallaby were introduced to the south island of New Zealand in the 1870s and within 40 years they numbered in the thousands; today there are approximately 15,000. So, why has the Rambouillet population in France gone nowhere? To answer this question, let's first examine the final stage of the invasion process, spread, in more detail by considering the two factors that primarily govern it: population growth and dispersal.

Growing populations

Population spread is principally a function of the established species' population growth. If conditions are suitable all populations will, at least initially, grow roughly exponentially. Since all organisms require some amount of physical space to survive, the addition of new individuals increases competition for resources. For motile species, such as most animals, increased competition will naturally lead some individuals to disperse to suitable unoccupied habitat. For sessile organisms, such as plants, only the progeny that land on suitable unoccupied habitat will survive while the remaining will either not establish or be crowded out before adulthood. In either case, the addition of new individuals into a resource competitive environment will ultimately result in the population expanding its geographical distribution. All else being equal then, a population with a high growth rate will spread more rapidly than one with a lower growth rate.

Mapping this spread geographically results in a roughly expanding circle that slowly envelopes more locations with time. In practice, the pattern is not perfectly circular because different habitats may be more or less suitable to the invading population, which changes the growth rate at the local level, increasing or decreasing the local rate of spread.

We see this pattern in the expansion of the common starling through North America after they were introduced into Central Park, New York, in 1890 by a local eccentric, Eugene Schieffelin

(Figure 6). Starlings evolved within the open grassland and shrub habitats of Eurasia and have most likely had a long history of inhabiting farmlands and pastures. Consequently, following initial introduction to the United States, the starling population expanded rapidly south and west across the grasslands and farms typical of the middle of the continent, and expanded more slowly across the high elevation forests of the Sierra Nevada and Rocky Mountains in the west. Their expansion north was also curtailed due to the cooler temperatures. Today, starlings continue to use grassy habitats typical of urban areas.

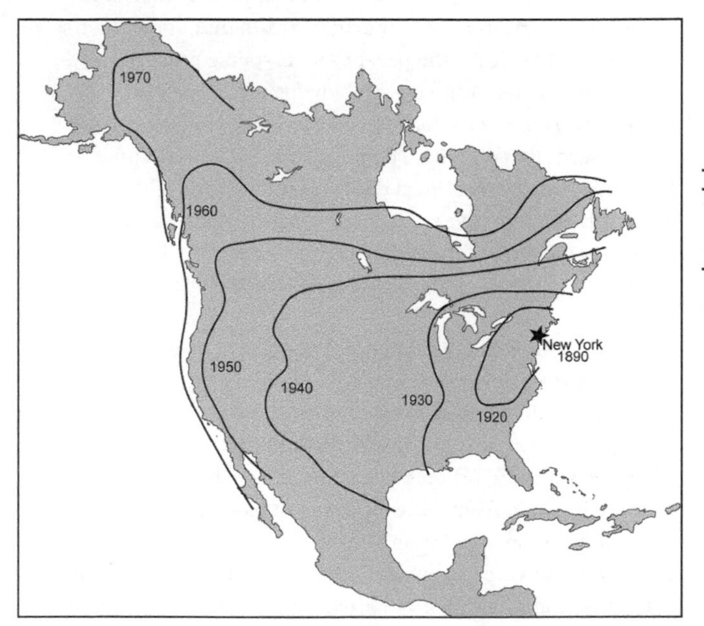

6. Geographical range expansion, or spread, of the common starling (*Sturnus vulgaris*) across North America since being introduced to New York City in 1890. The extent of range expansion through time is depicted with isolines. The further the lines are apart, the faster the rate of spread during that period, and vice versa.

Population dispersal

How far individuals disperse is the second primary factor that determines the expansion of an established species. Dispersal refers to an organism's ability to move from one suitable habitat to another and is principally a function of its evolution and the environment that it is attempting to move into and through. For animals, those species with longer or wider appendages associated with locomotion, such as wings or legs, will often disperse further than those without such traits. So, even if population growth rates were equivalent, giraffes would disperse farther than gerbils. Similarly, plants that exhibit traits that allow their seeds to be carried far from the parent will disperse more rapidly. Many children are familiar with blowing on dandelion (*Taraxacum sp.*) seeds, for instance, creating clouds of gently drifting seeds that each have their own parachute—a pappus—which allows the seed to travel great distances on the wind. Other plants have evolved alternative strategies to disperse their seeds, most notably in the form of fruits that are then eaten and distributed by animals.

These traits provide a fundamental rate of dispersal, a baseline, yet the greatest contribution to a population's expansion comes from those individuals that move unusually far from their location of origination (Figure 7). Consider a tamarisk tree whose seeds are dispersed via the wind much like dandelions. Most seeds will land near the invasion front—the location of the edge of the population furthest from the population's initial establishment location—and germinate. However, a few will be blown quite far and establish 'satellite populations': a non-native population far beyond the invasion front. These satellite populations aid expansion in two ways. First, they help colonize the intervening space between the main and satellite population as the invasion front now closes in from both sides. And, second, the satellite population can produce further satellite populations, resulting in a rapid expansion of the

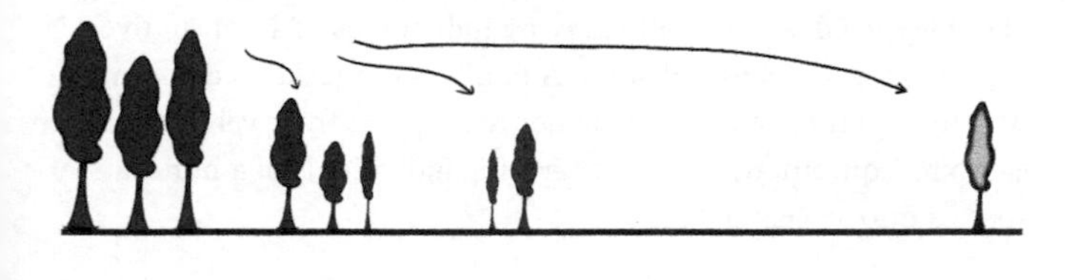

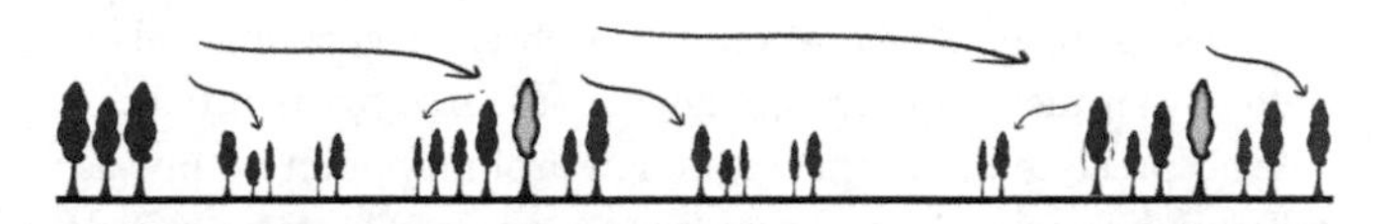

7. Simplified depiction of the expanding range front of a non-native tree population. Seeds from the adults are moved out into unoccupied habitat by the wind. Most seeds move a short distance from the stand, but a few are blown a long distance (grey trees).

invader across the landscape such that the invasion front jumps forward dramatically.

Satellite populations are not restricted to a handful of plants that benefit from gusty conditions. Non-native animals also establish satellite populations, typically due to human help. Here, for the second time during the invasion process, people are responsible for moving species from one locale to another. There are differences in the role humans play in the spread of established non-native species compared with the initial introduction of non-native species, with the most obvious being the distance and type of transport. During initial introduction, individuals of non-native species are generally carried over very long distances in ships or planes, while cars, trucks, recreational boats, and even bicycles provide the comparatively short-range and more frequent dispersal for already established species that are expanding their non-native ranges. Given our understanding of the introduction phase, government agencies can attempt to thwart invaders by examining imported goods; whereas it is difficult to stop members of society from accidentally moving individuals of a non-native species once it has established. A family on a weekend camping trip, for instance, will typically not realize that their vehicle or camping equipment could be carrying individuals of a non-native species into at-risk habitats.

Expanding, stabilizing, contracting

The dynamics of population growth and dispersal give rise to 'bi-phasic spread'. This is a commonly observed geographical pattern of invasive species spread characterized by an initial slow rate of spread that, at some seemingly arbitrary point in time, is followed by a rapid spread rate that is sustained for long periods. Bi-phasic spread was seen when emerald ash borer (*Agrilus planipennis*), a wood-boring beetle native to north-eastern Asia, invaded eastern North America in the early to mid-1990s. The emerald ash borer initially established in the United States near Detroit, Michigan. The initial population was small, highly localized, and exhibited some slow outward expansion across forests near Detroit. The initial spread rate was estimated to be about four kilometres per year until, in 2001, its rate of spread dramatically increased to 13 kilometres per year (Figure 8). The increased rate was due to numerous satellite populations establishing ahead of the invasion front leading up to 2001 and coalescing thereafter. As it happens, those satellite populations resulted from ash borer larvae being transported within firewood or ash trees that were used in commercial and residential settings. Emerald ash borer infestations have now been detected in 35 states across the USA.

Invasive species cannot expand their distributions indefinitely. Population expansion will ultimately be curtailed by a lack of suitable habitat in which to expand. For some species, such as the starling in North America, this might only happen after covering an entire continent (Figure 6). The starling spread eventually stopped when it reached the Pacific Ocean. Similarly, the very cold

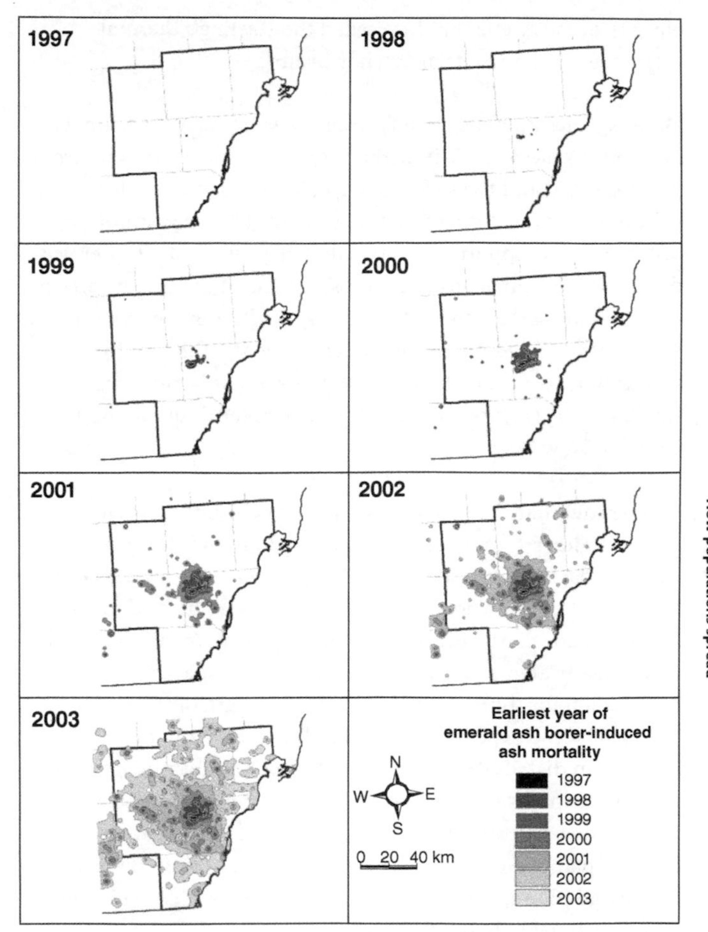

8. A map of the spread of emerald ash borer from its initial point of establishment in Detroit, Michigan.

climate of the Arctic Circle exceeded the starling's thermal tolerance and halted its march northward.

Other species do not spread far from their initial establishment location. In these cases, it might be that suitable conditions are very localized and the surrounding habitat too unsuitable to sustain the population, or that local conditions may simply be insufficient to support a high population growth rate. Either way, the invader remains only locally distributed. What might come as a surprise though, is that a failure to spread is quite common among established non-native species. About a third of all successfully established non-native birds have distributions smaller than 150 km^2, an area not much larger than Manhattan Island in New York.

Having said that, we cannot assume that non-native species that appear to have remained localized will do so indefinitely. Invasion-lags refer to several related phenomena that can occur throughout the invasion process. In the spread phase of invasion, lag events are characterized by, effectively, no population growth and no associated expansion of the non-native species' distribution for some seemingly arbitrary length of time, followed by the expected rapid population growth and rapid increase in distribution. This phenomenon is not uncommon and has been observed in several groups of invasive species. Sami Aikio and colleagues examined 105 of New Zealand's invasive plants and found nearly all of them exhibited a lag in the spread phase, with the average time between establishment and spread being between 20 and 30 years. Similarly, Kevin Aagaard and Julie Lockwood examined 17 invasive bird species present in Hawaii and found approximately 80 per cent exhibited lag times ranging between 10 and 38 years. Although these and many other studies document invasion-lags, the underlying causative processes are poorly understood, rendering prediction of when species will spread and which will not intractable.

Where invasion-lag, or specifically spread-lag, has a non-native population remaining in stasis for an arbitrary length of time before exploding in numbers, the reverse is also true; a well-established and widespread non-native species can rapidly regress in distribution and numbers, or disappear entirely. However, beyond situations where our control efforts manage to supress or eradicate an invasive species, it is not always clear why or how this happens. These 'boom-and-bust' episodes do appear more common on islands, or with species that experience high variability in population size that serves to rapidly increase numbers which subsequently collapse naturally.

There are many reasons why established populations may collapse. One challenge for invasive species is low genetic diversity due to the limited number of founding individuals that initially established the population. Low genetic diversity reduces the population's ability to adapt to new environments or endure when conditions are unfavourable. Another factor is 'enemy accumulation'. You will recall that introduced species can 'escape' their native range enemies, which aids their initial establishment. Without that ecological pressure the population can grow and expand rapidly. Over time though, the invader may accumulate new enemies and when those enemies begin to exploit the booming invasive population, individual survivorship and reproductive rates drop, sometimes precipitously, and the invasion may collapse.

This latter scenario appears to have played out with one of the world's most ecologically and economically damaging invaders, the yellow crazy ant (*Anoplolepis gracilipes*). It is native to south-eastern Asia but has established numerous non-native populations around the world, including in the Seychelles Islands off Africa, on Christmas Island, and in Arnhem Land, Australia. Several colonies of the ant have been observed to reach very high abundances, only to decline without management intervention. The supercolony on Christmas Island for instance, which reached more than 20 million ants per hectare— that is 2,000 ants per square metre—declined over 18 months to a level that allowed native fauna to recolonize sites that were previously made uninhabitable by the ant's presence. And in Arnhem Land, several yellow crazy ant populations declined with three colonies going extinct, including a well-established colony that occupied roughly four hectares of savannah woodland. The reason for these collapses, at least in Arnhem Land, appears to be due to pathogens that reduced reproduction rates in the colony and ultimately led to its demise.

We now return to where we started and ask, why has Bennett's wallaby in Rambouillet forest not spread across France and become invasive? Although we haven't discovered the specific reason in this case, you now know the general conditions. The wallaby's population growth must not be sufficiently high to cause it to spread to other areas—at least not yet. And you can add to this that its population growth is not sufficiently high because either the environment is not well suited, or the surrounding environment into which it could spread is not suitable. Either way, the population is small enough and impact slight enough in Rambouillet that they are not yet considered 'invasive'. Still, there are many other examples of non-native species that grow in numbers and spread across landscapes, and it is in their wake that the impacts follow.

Chapter 6
Interactions in ecosystems

Recall that an invasive species is defined by two characteristics. First, they are not native to a region, and second, they cause undesirable impacts or have spread beyond their initial introduction location. We have hitherto explored how a non-native species can establish a population and spread across a novel landscape. Now we turn our attention to how non-native species cause undesirable impacts. To do this, however, we need to understand a little about ecological interactions and, in the following chapter, ecosystem states.

An ecosystem is defined as the community of organisms that exist within a given area that interact with one another and the physical environment. A grassland in Britain is an ecosystem, as is a lake in North America, a desert in Africa, and the freezing waters of the Antarctic. Although the most obvious features that come to mind when thinking of an ecosystem are the plants and animals, especially the gregarious species, it is the ecological interactions that are the constant as they are the fundamental processes that shape the ecosystem. Take predatory interactions, for example. Regardless of the species present, for the most part, all ecosystems will contain animals eating other animals, animals eating plants, and in some cases, plants eating animals. There are numerous ecological interactions ranging from the subtle to the obvious in severity of effect. As we cannot provide an exhaustive list here, we will focus on the major types of interactions that invasive species have in ecosystems.

Eating the natives

Consuming other species in an ecosystem is one of the most obvious and impactful interactions between species. All heterotrophs—species that rely on others as a source of food—do this regardless of whether they are top-order predators or herbivores. This interaction can be especially problematic if the non-native species is doing the eating since the native prey and non-native predator do not often share an evolutionary history. Without that shared history, native species are often behaviourally naive to the predator, which makes them easier to catch or consume. Similarly, without a shared evolutionary history, native plants may not have appropriate defences against non-native herbivores.

Non-native herbivores have considerable impact on ecosystems because they feed at the base of the food web. Some herbivores, regardless of whether they are native or invasive, may consume all or most of the plant, whereas others only consume portions such as the leaves, fruits, roots, or inner tissues like the phloem. If a herbivore's impact on important vegetation communities is sufficient, they can cause a trophic cascade that affects many other species not directly involved in the herbivory, or even cause physical changes to the ecosystem itself. For this reason, goats, pigs, horses, and rabbits have historically been among the most destructive invasive herbivores. On Laysan Island in the Hawaiian Archipelago, for instance, rabbits (*Oryctolagus cuniculus*) eliminated 26 species of native plants after having been established on the island for only 20 years. Because these native plants provided critical animal habitat, the loss of vegetation contributed to the loss of three native bird species and five species of insect. Similarly, in Australia, invasive rabbit populations caused huge swathes of erosion when they burrowed and over-grazed vegetation that held soils together, which in some instances has rendered agricultural lands useless.

In recent decades, non-native herbivorous insects have become one of the most economically and ecologically harmful species. Consider just one group, invasive foliage-feeding insects. When in sufficient abundance they can completely defoliate a tree within the span of a year. Since the leaves are the site of photosynthesis, any tree that loses a sizeable portion of its leaves will develop more slowly, reduce reproductive output, and potentially die.

Most foliage-feeding insects restrict their feeding to a taxonomic group of trees, making their impacts largely species-specific. The spongy moth (*Lymantria dispar*), which was introduced into North America from Europe in 1869 to develop a silkworm industry, is a classic and disastrous example. Spongy moth larvae—caterpillars—consume the leaves of several native North American tree species, but predominantly native oaks (*Quercus* species). The caterpillars will on occasion kill an oak tree outright, but most often their feeding results in reduced seed output and tree growth. Over time this has led to forest-wide shifts in tree communities where oak, the once dominant native tree, has been replaced by other tree species.

Predatory species, especially top-order predators, have consistently been among the most impactful invasive species in natural ecosystems. The impact can be particularly severe when the invader exhibits novel hunting strategies in the ecosystem, as is the case with Burmese pythons in the Florida Everglades. Burmese pythons are a large constrictor native to south-eastern Asia and have been established in the Everglades since the early 1990s. Although there are other snakes in the Everglades, the ecosystem has not had a large constricting species living within it for at least 15 million years. Couple this history with the fact that Burmese pythons grow to tremendous sizes, five metres in some cases, then there is little wonder that they have had a devastating impact on the ecosystem. In locations where pythons are present, mammal surveys have documented declines of 90 per cent or more in native rabbits (*Urocyon cinereoargenteus*), foxes (*Vulpes vulpes*), raccoons (*Procyon lotor*), opossum (*Didelphis virginiana*), white-tailed deer (*Odocoileus virginianus*), and bobcat (*Lynx rufus*) populations. Where Burmese pythons are absent, these species are widespread. In predatory interactions, a lot can be lost from ecosystems rather quickly.

Competing with the natives

A non-native species entering an ecosystem will almost certainly compete with several native species for resources such as food, light, nutrients, or the most basic resource, space. In the extreme, if the invader is a better competitor, they can exclude the incumbent native individuals entirely from those resources, resulting in the latter's death or population extirpation. The sometimes extreme physical and financial efforts that farmers expend to clear their crops of non-native plants, commonly referred to as weeds, highlight just how severe and obvious an interaction competition can be. On the other hand, the effect of competition can be subtle when interactions do not directly result in death of the individuals involved and require several generations for effects to become apparent. This is often the case where competition merely reduces the native species' reproduction and growth rates.

Since photosynthetic plants require sunlight for survival, competition for light is a principal way that non-native plants affect other plants. Of course, sunlight is not in short supply, but access to it can be. Plants have therefore evolved a host of traits to compete for light that, while in their native community, provide a sufficient level of success against co-evolved strategies. The problem comes when a non-native plant has a dominant strategy that native plants cannot counter. A well-documented example of this comes from the invasion of Britain by giant hogweed

(*Heracleum mantegazzianum*). Hogweed is a herbaceous flowering plant native to eastern Europe that was introduced to Britain, and other locations, as an ornamental. It was a particularly popular garden plant in the 19th century. Once it escapes the garden, it prefers to grow in grassy areas where it forms large, dense stands that shade out native grasses and flowering plants, and in some locations where invasive populations of giant hogweed are present, up to 20 native plant species have disappeared.

For sessile and semi-sessile species, competition for space is common. Consider *Dreissena polymorpha*, the zebra mussel, which was introduced into the North American Great Lakes from Europe in the 1980s via ship ballast water. Since all mussels spend much of their life attached to hard and stable surfaces, zebra mussels directly compete for space with native North American *Unionid* mussel species. Unfortunately for the natives, zebra mussels are better competitors. Zebra mussel populations grow quickly as individuals reach maturity and reproduce after only a year, compared with the three to five years that native mussels require. As the zebra mussel population grows, they occupy every available hard substrate and, when that is not available, they settle directly on native mussels, effectively smothering them and restricting the native mussel's ability to feed and grow. At high densities, smothering is only one aspect of the competition, as zebra mussels also reduce nutrient resources available in the water column. For native mussels, then, not only are there fewer locations available in which to settle and collect food, there is also much less food in the water, both of which are detrimental to the native species.

Competition for space also occurs between mobile species. Let's return to the yellow crazy ant on Christmas Island. The island is famous for the endemic red crab (*Gecarcoidea natalis*). There were once 100 million red crabs on the island and their feeding activities—moving seeds and fruit to their burrows and preferentially consuming some saplings over others—made them a primary engineer of the forest ecosystem. The introduction of the crazy ant altered this dynamic for the decades when they were superabundant on the island (see above for details on their decline). Red crabs do not eat or harm the ants, but when crabs are on the move crazy ants squirt formic acid in misguided defence, which ultimately leads to the death of the crabs. Couple this impact with competition for burrow space, and crazy ants have killed tens of millions of red crabs, which has resulted in a cascade of ecosystem-wide effects. The non-native stinging tree (*Dendrocnide peltata*), for instance, has increased in numbers; increased leaf litter and saplings in the understorey benefited another invader, the giant African land snail (*Achatina fulica*), while resulting in a concomitant decrease in the native island thrush (*Turdus poliocephalus*), emerald dove (*Chalcophaps indica*), and Christmas Island gecko (*Lepidodactylus listeri*). Competing for space, or any resource, while perhaps not as immediate as the effects seen in predatory interactions, can still result in invasive species causing dramatic ecosystem-wide impacts.

Infecting the natives

When a species invades a habitat, they potentially bring parasites and pathogens with them that can affect native species beyond the competitive or predatory interactions noted. Parasites and pathogens are organisms that require a host species for at least one stage of their life cycle. The host provides a critical resource, often nutrition, which allows the parasite or pathogen to successfully breed, grow, or metamorphose into a new life stage. The cost to the host can range from an annoyance, to reduction in growth or reproductive rates, to, in the extreme, death. These costs typically vary in intensity depending on whether the host experiences other stressors such as a lack of food, or whether the host has evolved defences against the parasite or pathogen. Some parasites and pathogens are restricted to a single host and, therefore, if introduced with their host to a novel habitat, present little concern for native species; in fact, they may reduce the likelihood that the invasion will succeed. Many parasites and pathogens are nevertheless not restricted in this way.

The Pacific oyster (*Magallana gigas*) is a commonly farmed shellfish native to coastlines of the North-West Pacific Ocean and Sea of Japan, and provides a cogent example of the impacts of an invasive parasite species. The culinary popularity of the oyster led to its importation and establishment over much of the world for aquaculture, but as it often escapes, it is now also one of the most widespread marine invaders. Unfortunately, the oyster does not always travel alone and its parasitic copepod, *Mytilicola orientalis*, has been co-introduced with it. In the Wadden Sea, the invasive parasitic copepod has found new hosts in the native blue mussels (*Mytilus edulis*), common cockles (*Cerastoderma edule*), and Baltic tellin (*Limecoma balthica*). The invasive copepod resides in the intestines of native mussels during its juvenile stage, consuming its host's intestinal tissue as well as nutrients that flow through its gut. This slows the host mussel's growth and increases the chance of its death. Parasitism of native blue mussels by the invasive copepod has increased in recent decades and is causing both economic and environmental impacts, since blue mussels are commonly harvested for food and they play a key role in maintaining water quality in the Wadden Sea.

A big impact from non-native pathogens and parasites is in the diseases they cause. In plants, more than half of emerging infectious diseases are caused by non-native pathogens, and non-native fungi or fungal-like infections are the main cause of emerging infectious diseases among forest trees. These invasive pathogens include the infamous chestnut blight and Dutch elm disease, the latter of which over the 20th century caused most mature native elm trees to disappear from Britain and continental Europe.

Animals have not escaped disease-causing invasive parasites and pathogens either. The global decline and extinction of native amphibian populations has, to a large extent, been caused by chytridiomycosis, an infectious disease caused by the invasive chytrid fungi *Batrachochytrium salamandrivorans* and *B. dendrobatidis*. With an entire group of species disappearing from an ecosystem, cascade effects naturally follow. One study found that after widespread amphibian loss due to chytrid fungus tropical snake diversity in Panama had also collapsed. It is no surprise then that a quarter of the '100 worst invasive alien [non-native] species', documented by the International Union for the Conservation of Nature, are non-native parasites and pathogens.

Physical interactions

Thus far we have considered interspecific interactions, interactions that occur between members of two (or more) species. A final set of important ecological interactions is between the invader and their surrounding physical environment. An ecosystem's abiotic components include its chemical attributes, such as the pH of water, through to its topography and structure, which directs water and nutrients through the terrain or provides shelter for organisms. By affecting the physical characteristics of the environment, invasive species can substantially affect the niche space available for other species.

Returning to giant hogweed, we see how an invasive species can tip that first domino, sending change toppling through an entire ecosystem. Ordinarily, streambanks are relatively robust as species-rich native plant communities hold the soil together and prevent erosion. But when hogweed establishes along river margins, it crowds out the natives and becomes the only plant growing there. Since hogweed does not have the same extensive root systems as native plants, soils underneath hogweed stands are not bound tightly. Thus, when rainstorms occur, streambanks tend to erode much faster under hogweed patches. This high rate of erosion increases the amount of sediment in the stream, which increases water turbidity and results in less light for stream-dwelling plants. Reduced light reduces the suitable habitat for submerged plants and with fewer plants in the stream, there are fewer places for juvenile animals to shelter, and water flow increases. Increased water flow then causes further erosion that can shift the course of streams and result in significant impacts on the availability and timing of downstream water flow. This cascade of effects from a single invasive species is not unique to hogweed. A similar scenario plays out with feral horses (*Equus caballus*) in the Australian Alps; being a heavy bodied, hard hooved animal, horses trample stream bank edges and cause them to collapse, and the dominos fall.

Looking to non-native earthworms we see that even the smallest, and in this case often unseen, invasive animal can affect the physical properties of entire ecosystems. Non-native earthworms have been introduced to North America over the last few centuries from Europe, in the case of *Lumbricus* species, and Asia, in the case of *Amynthas* species. This introduction has been quite problematic since there were no native earthworms in many temperate and boreal North American forests and therefore no shared evolutionary history between the worms and the native species that call these forests home. As earthworms move through soils, they mix the organic and inorganic soil layers. This action reduces the concentrations of carbon and nitrogen in these soils, and in some locations completely removes the organic layer. This mixing naturally reduces the suitability of those soils for shallow rooted native plants, which are common in the forest understorey, and invasive earthworms have ultimately led to decreases in species diversity of herbaceous plants in several ecosystems including Chippewa and Chequamegon National Forests in the United States.

Some species directly affect the chemical composition of an ecosystem, which can affect the survival and reproduction of other species. Allelopathic plants, for instance, deposit biochemicals into the environment that suppress their would-be plant competitors. The compounds are deposited in several ways but in plants it is often by direct exudation from the roots or from foliage that falls to the ground and decays. *Casuarina* spp., often called Australian pine, is in this latter group. They are native throughout south-eastern Asia and north-eastern Australia and are now invasive in the United States, South Africa, India, and Brazil. Depending on the soil, the tree reaches 6 to 30 metres tall and as it does it drops its foliage—twigs covered in scale leaves—at the tree base. Both the live tree and the decaying foliage release phytotoxins, specifically phenols, which increase soil acidity and subsequently suppress the growth of other plants around the base of each tree. Where *Casuarina* has invaded, native understorey plants do not often grow.

The direct impact from *Casuarina* is a loss of or reduction in the native understorey plant community; but there are secondary effects too, including a wholesale change in vegetation structure. Where *Casuarina* trees supplant other trees and suppress the growth of understorey plants, the resultant monoculture exhibits very little structural diversity. That is, the canopy of the ecosystem will be much the same height across the affected area, and there will be few if any levels between the canopy and the forest floor that native fauna would ordinarily feed on or shelter within. For instance, native browsing herbivores, which are limited by both the species of plant they can eat and the height that they can reach, would naturally need to find a new *Casuarina*-free habitat with suitable plants and structure. Similarly, small foraging species that might ordinarily use forest understorey as shelter to escape predation might experience a rapid increase in predation within *Casuarina* forests until they too find an alternative habitat.

A final important effect to highlight occurs when invasive species affect disturbance regimes such as fire or flood. Fire is a particularly strong structuring force of most ecosystems.

Ecosystems that are naturally fire-prone are populated by native plants and animals that either tolerate fire or require fire to complete their life cycle: for instance, species of *Banksia* and *Eucalyptus* require the high temperatures that are only achieved during wildfire to trigger seed release. Key to sustaining this evolutionary relationship is the periodicity with which fires occur, how hot the fire burns, termed fire intensity, and what plant parts are consumed. Altering any of these factors by too great a degree will negatively affect pyrophytic plants and therefore the ecosystem more broadly. As an example, in a *Banksia* woodland in Western Australia, high-intensity wildfires reduce adult plant survival and seed production, which ultimately decreases habitat suitability for the endangered Carnaby's cockatoo (*Zanda latirostris*).

Grasses are one group of species that can have a dramatic effect on fire regimes. In Australia, the invasion of semi-arid ecosystems by buffel grass (*Cenchrus ciliaris*) markedly shortens intervals between fires and increases their intensity. This effect is due to the invasive grass growing more rapidly than native species, which increases 'fuel loads'—vegetation that fuels fire—more rapidly than normal. In fact, this process can create a positive feedback loop that aids in buffel grass expansion. As fire frequency and burn intensity increase, native plants decrease and the invasive buffel grass expands into those patches, leading to further increases in fire intensity and frequency, further decreasing native species. Native plants in the ground layer are most affected initially, but eventually fire frequency and intensity is sufficient to affect overstorey species too, such as the iconic fork-leaved corkwood tree (*Hakea divaricate*).

Non-native pine trees (*Pinus* species), among others, lead to a similar and particularly pernicious cycle in the fynbos, the shrublands and heathlands of South Africa. Like buffel grass, invasive pine trees in the fynbos increase fuel loads beyond that of the load native trees generate, despite it being a naturally fire-prone ecosystem. And again, the increased fuel load increases fire intensity and frequency, which reduces native plant abundance. Invasive pine trees further suppress native species through allelopathy and by outcompeting them for sunlight. The final blow to natives comes in the form of reduced water supply because, first, the invasive pines consume more water than native shrubs, and second, the change in understorey vegetation caused by the invasive pines allows more water to run off, reducing stream flows that then result in a drier environment and further compounding the effects of fire. As an aside, one study estimated that reduction in stream flows in South Africa due to black wattle (*Acacia mearnsii*), another invasive tree, has cost US$1.4 billion since introduction. This ecological feedback relationship between invasive pines, fire frequency, and water flow has led to large-scale transitions in the fynbos ecosystem that hinder the many endemic species in the region.

As should be apparent at this point, especially from this last example, ecological interactions do not occur in isolation. Any given species will exact a multitude of ecological interactions on the abiotic and biotic environment around them. Some of these processes will be amplified, in positive feedback loops, as in the fynbos example above, while other processes might be dampened in negative feedback loops, such that any potential impact is reduced. The problem of non-native species is that because they often do not share an evolutionary history with their novel environment, the effect of their ecological interactions can propagate through the environment as the dynamic state of the ecosystem absorbs the new species. And it is through these shifts in ecosystem states that invasive species have the greatest impact.

Chapter 7
States of an ecosystem

Ecological interactions, along with the abiotic and biotic components of the environment, determine an ecosystem's 'state'. The concept of an ecosystem state reflects the fact that ecosystems are not static entities; they are dynamic and respond to their inputs. Individually, we are most familiar with short-term changes. Consider the annual cycle of a deciduous forest. Trees lose their green, blush in brilliant oranges and reds, then stand bare until reawakened by spring; during which time animal populations wax and wane as they migrate, reproduce, and respond to changes around them. And this is just one breath of the forest. Other cycles at shorter and longer time scales exist too. A little extra rain this year leads to an increase in herbs in the understorey in the next, the extra vegetation results in a bumper reproduction season for a grazing herbivore, which spurs growth in the predator population. With close enough attention, we need not leave our backyard to witness these transformations.

By lengthening our time horizon, these short-term cycles are wiggles in the long-term state of an ecosystem. The Sahara is the archetypal image that springs to mind when we think of a desert: rolling sand dunes straddling horizons on all sides with unyielding sun blazing down from above. But this is just one of many states in which that ecosystem could exist. About 10,000 years ago, the region was wetter and a large portion of what is now hot dry desert was covered in moist tropical woodlands and grasslands. That was another state. Ecosystems exist within a particular state while the physical conditions and composition of their biotic communities are roughly constant. But, when the size or duration of an impact becomes too great—for instance, from the impact from invasive species—ecosystems can and will shift to a new state, perhaps irreversibly.

Ecosystem states

A ball-and-hill analogy is useful for envisioning ecosystem states and how invasive species might influence the state of an ecosystem. In Figure 9 the X-axis represents a continuum of states in which the ecosystem, the ball, could exist. The line on which the ball rests is the 'curve of stability'. The curve traces out valleys, places of stability, slopes, states characterized by instability, and peaks, unstable states that represent tipping points between stable states. Where gravity will cause a ball to roll down a slope to a valley, in our analogy ecological feedback processes take the place of gravity, nudging ecosystems into more stable states. With this analogy, then, we can imagine the 'effort' required to push the ball in Figure 9 out of some state, say, C, and into another state, B.

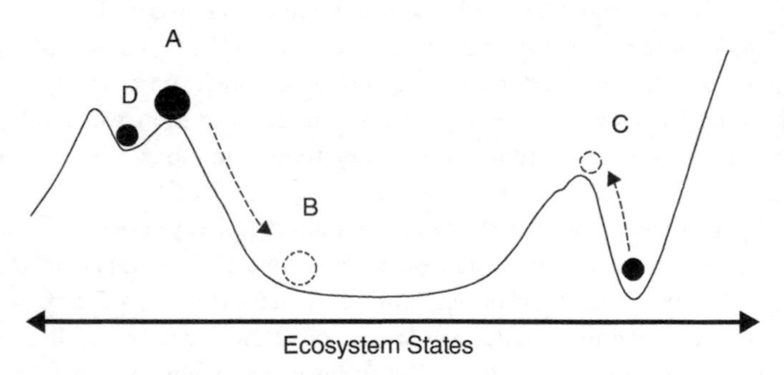

9. Representation of the stability of an ecosystem in various states. Peaks represent unstable states and ecosystems would most likely transition quickly through such states to valleys, stable states.

To achieve this the system would need to be changed enough, nudged enough, to get the ball out of the valley and to the top of the peak, a tipping point, where once the ecosystem crosses that threshold it transitions to the new state with no further input.

The concept of ecosystem states should be familiar to anyone with a home vegetable garden. The garden is a small ecosystem that the grower attempts to keep in a specific state, namely the maximization of fruit and vegetable production. To achieve this, the grower is almost always intervening in the dynamics of the ecosystem; they remove unwanted plants that begin to grow and perhaps spray insecticides and fence off the patch to stop insects and other animals from consuming the vegetables. Since maximizing vegetable growth is an inherently unstable state for the ecosystem, in our analogy, the grower is effectively keeping the ball on a slope. If the grower stops intervening, even for a day, the ecosystem, that small patch of ground, will naturally begin to shift to a more stable state. Vegetables may still grow, but yield will almost certainly be lower as other plants crowd out the vegetables and wildlife consume the produce.

Ecosystems can also exist in locally stable but globally unstable states. Let's say that the state at position B in Figure 9 represents pristine forest. The wide valley represents a continuum of states with similar relative stability where the ecosystem wanders as plant and animal communities wax and wane with long-term environmental cycles. Now, let's imagine a major disturbance event, perhaps a one-in-a-hundred-year wildfire, pushing the ecosystem towards a new state, position A. If the fire is not sufficiently severe and the ball only reaches part way up the slope towards A, then the ecosystem will naturally return to the stable valley of position B; that is, the forest will regenerate. But, perhaps a one-in-a-thousand-year fire is so severe that it kills many or all of the old overstorey trees and their seed bank; then the ecosystem may be pushed beyond the tipping point of A and quickly transition to D. This new locally stable state might be characterized as a shrubland, where shrubs and other small ground covering plants exclude nearby encroachment from large forest trees. However, because the fundamental abiotic conditions have not changed, this locally stable state is globally unstable, and so a relatively small disturbance event, or the introduction of a non-native species, may push the ecosystem out of this small stable valley; perhaps towards position B, or perhaps further away from what it once was.

It is changes to an ecosystem's abiotic or biotic properties that drive it into new states. Changing the composition of species or number of individuals within a species is akin to nudging the ball in our analogy left or right. If it's in a deep valley, more effort will be needed to push the ball up the hill before reaching another stable state than if in a shallow valley. Similarly, a resilient ecosystem will need to see greater change in its species composition to shift it to a different state. Abiotic aspects of the system such as rainfall, nitrogen availability, or even the ecosystem's topography affect the shape of the stability curve itself (Figure 10). A shift in rainfall patterns, for instance, may result in a previously stable state becoming unstable and consequently cause the ecosystem to shift to a new stable state, perhaps one that was once unstable.

The abiotic and biotic changes obviously do not occur in isolation. There are constant positive and negative feedback processes between those domains. Changing the physical properties changes the composition of species in a system since those physical properties affect the niche space available to organisms; but changes to the species composition can affect the ecosystem's physical properties too. Recall that hogweed affects the stability of stream banks where it crowds out native species. This complex and dynamic nature of ecosystems means that the introduction of a non-native species immediately begins to shift the ecosystem's state. Many ecosystems are sufficiently resilient that the presence of the non-native species, at least initially, does not alter the state

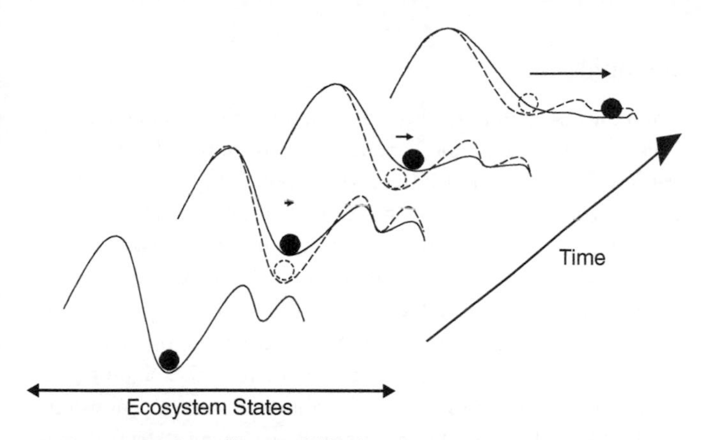

10. Representation of how stability in ecosystem states can change through time as physical properties of the ecosystem are altered, such as rainfall or nitrogen availability.

of the ecosystem. But when the introduced species reaches high population densities the chance of them causing the ecosystem to transition to a new state is much more likely.

Consider North American beavers, which were introduced to Tierra del Fuego in South America in the 1940s for the fur trade. They now number in the hundreds of thousands and efforts are under way to remove them since they have dramatically affected the state of the native wetland ecosystem. Many species of tree in Tierra del Fuego do not readily regrow after beavers gnaw them. This has left parts of once pristine forested landscapes barren. But the stability curve of these ecosystems has changed shape too as water and nutrient cycles have been affected by beaver dams. With new niches available, many non-native plants have managed to establish in the region, adding to pressures and pushing the ecosystem into new states, further away from what it once was. Returning the wetlands of Tierra del Fuego to what they once were may not be practicable or even possible, although there are massive efforts under way to try to do so.

Bistability

From the above ball-and-hill analogy, it might seem that if an ecosystem can transition from one state to another then surely it can shift back; or to put it in more technical language, ecosystems states are continuous. Ecosystem states certainly can be continuous. Figure 11(a) plots a continuous set of ecosystem states as a function of an environmental variable. The example we use here is water clarity in a shallow freshwater lake as a function of nutrient load. When nutrient load is relatively low the lake is clear, yet as nutrient load increases the ecosystem reaches tipping point 1 and rapidly transitions to a new stable state, one of poor water clarity. Since this is a set of continuous ecosystem states, the reverse transition is also possible. As nutrient load reduces the lake will naturally reach tipping point 2 and rapidly transition from poor to good water clarity.

Alternatively, ecosystems can transition discontinuously, as shown in Figure 11(b). Here, the transition from high water

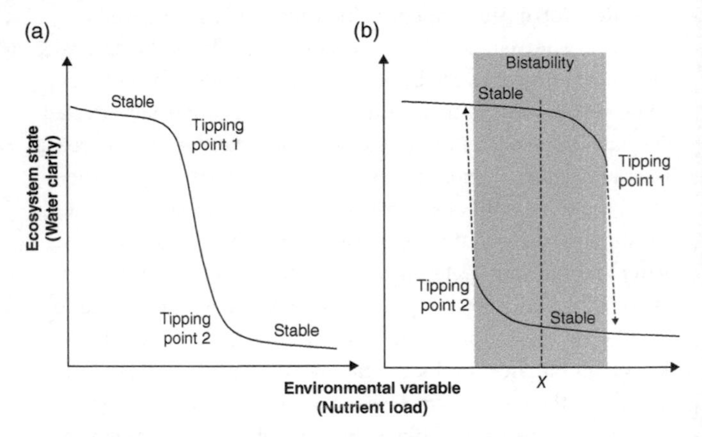

11. Ecosystem states can vary (a) continuously or (b) discontinuously. This difference can lead to bistable ecosystem states.

clarity to low water clarity occurs abruptly at a different level of the environmental variable than does the transition from low water clarity to high water clarity, a condition known as hysteresis. This results in a range of values that the environmental variable can take where the ecosystem can be in one of two stable states, the region of bistability. If the nutrient load of the system was X in Figure 11(b), for instance, then the lake could exhibit high water clarity or low water clarity depending upon which tipping point the system had most recently transitioned to.

In theory, bistable ecosystems should transition between their states when the appropriate tipping point for a given environmentally important variable is reached. But in practice such transitions may not occur, resulting in what is generally referred to as ecosystem collapse. Sticking with the shallow freshwater lake example, the ecosystem could transition from a state of good to one of poor water clarity as nutrient levels increase. But, if tipping point 2 in Figure 11(b) occurs at a biologically impossible nutrient level, then the reverse transition will not occur naturally (at least not on important socio-environmental time scales). In such cases, the only way for the ecosystem to regain the state of good water clarity is through external intervention, often by land managers conducting ecosystem restoration efforts.

The effect of invasive species on bistable systems complicates the above discussed dynamics further. In a global meta-analysis on the impacts invasive species have in shallow freshwater lakes, Reynolds and Aldridge found invasive fish and invasive crustaceans can lead to early ecosystem collapse, and delay or even block recovery. This occurs because of the impacts that invasive fish and crustaceans have on aquatic plant densities, nutrient levels, and phytoplankton generally, all of which affect water clarity in various ways. The resultant effect is a shift in the region of bistability, and the associated tipping points, to a lower nutrient level (Figure 12(a)). In some scenarios, where level *X* in Figure 12(a) marks the minimum biologically possible nutrient level, then tipping point 2 may not be reached naturally while invasive fish and crustaceans are present.

On the other hand, Figure 12(b) shows the shift in tipping points caused by invasive molluscs. Since molluscs filter water during feeding, they reduce phytoplankton populations directly and improve water clarity. This outcome seems like a benefit, and indeed it is within already degraded ecosystems as it promotes early recovery, but there is a downside beyond the effects invasive molluscs have on other species: they mask the impact of eutrophication. If nutrient levels in a lake are increasing due to farming run-off, let's say to level *Y* in Figure 12(b), invasive molluscs may make an ecosystem appear to be in good health. Yet, removal of invasive molluscs at this point then results in the ecosystem transitioning to poor water clarity and additionally shifts tipping point 2 back to the original level, making recovery more difficult.

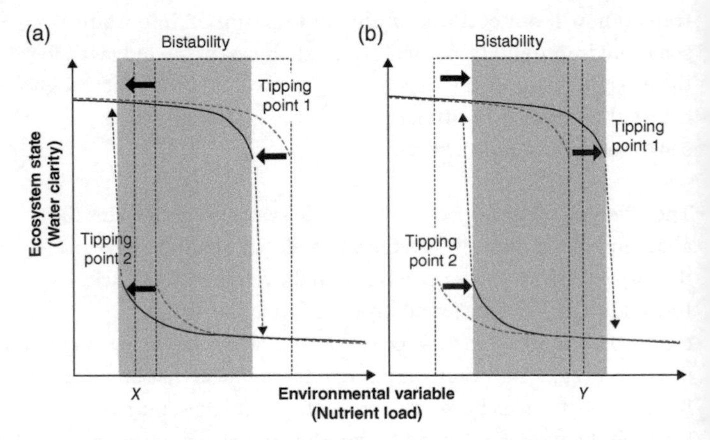

12. Invasive species can shift the tipping point in bistable systems, making ecosystems (a) more likely to collapse and delay recovery, or (b) delaying collapse and promoting an early recovery.

Bistability, or in reality multistability, has been observed in other ecosystems too, grasslands in deserts for instance. While those cases present different examples, similar processes and complex dynamics are at play: ecosystems exist in a range of states and transitions from one to another may not be reversible. Where the abiotic and biotic community share a long evolutionary history, ecosystems have had sufficient time to settle into stable states. Transitioning from this state is normally slow, as environmental conditions and cycles are also typically slow. And where the transition is rapid it often follows from a considerable perturbation, a major storm or fire event, for instance.

Non-native species upset this balance. They shift the dynamics, amplifying some feedback processes and dampening others. The ecosystem responds by 'seeking' out a new stable state. If the ecosystem is sufficiently resilient or the non-native species has minimal impact, the new state might be similar to the original. Indeed, the wallabies of Rambouillet forest are yet to bring about major change. But it is clear some ecosystems are affected much more or are already under pressure due to other human caused impacts. So, the addition of a non-native species is enough to dramatically affect the ecosystem, like the invasive comb jelly of the Black Sea that contributed to the collapse of an important industry. The ecological interactions caused by non-native species are obvious, they are the brown dots on the skin of an apple infected by fruit fly, but the impact invasive species have on the overall state of the ecosystem is the rot in the apple's core.

Chapter 8
Undesirable impacts

In the last two chapters we examined *how* invasive species cause undesirable impacts, namely through their ecological interactions with the abiotic and biotic environment and, ultimately, by affecting an ecosystem's state. Here we take a closer look at what those undesirable impacts are and clarify why such impacts are undesirable, though many are self-evident. To do this, we will categorize undesirable impacts into economic, human health and wellbeing, and environmental impacts. These categories are useful for recognizing how impacts are measured and recognizing the different values at play in determining whether an impact is undesirable.

Economic

The types of invasive species impacts that are often categorized as economic range from the obvious, such as reductions in agricultural and farming yields or infrastructure and property damage, to the more abstract, the loss of opportunities or loss of ecosystem services. We need not spend much time on the obvious examples. We have seen how invasive insects, specifically fruit fly, can reduce farming yields. And many readers are familiar with how invasive species can damage property. Formosan termites (*Coptotermes formosanus*), which are endemic to southern China but have been introduced almost worldwide, cause more than

US$1 billion in damages annually to houses and other wooden structures in the United States alone. But it is worth briefly examining how what might seem to be solely an environmental impact can be a major economic one.

Ecosystem services are the things society gets 'for free' from living within or near functioning ecosystems. These services include production, regulation, and even culturally important services. Production services refer to the food, fibre, and fuel generated by ecosystems, as in when wood is selectively harvested from a naturally regenerating forest for paper products and building materials. Regulatory services refer to those processes that regulate the environment such as carbon capture and storage by plants, capture and purification of water, and the retention and slow release of agriculturally important nutrients like nitrogen and phosphorus. Cultural services include a wide variety of positive benefits of nature to society. These benefits might be tangible recreational activities, the use of a reef system by scuba divers, for instance, to the more intangible, such as a sense of wellbeing, spiritual inspiration, or the cultural heritage that one derives from an ecosystem.

Although ecosystem services are provided by ecosystems 'for free' they still have an economic value to society, which can be estimated. Advances in environmental economic methods over the last 20 years have generated numerous estimates for both total value and biome-specific ecosystem services. Robert Costanza and colleagues estimated that in 2011, total global ecosystem services were worth between $125 and $145 trillion per year, which was about 30 per cent more than the combined gross domestic product of all nations at the time. Given the size of this number, perhaps a more appreciable figure can be found in estimates by Rudolf de Groot and colleagues. They reviewed more than 300 publications and found the total value of ecosystem services varied between biomes: terrestrial biomes ranged between a mean of US$1,588 per hectare per year for woodlands and US$5,264 per hectare per year for tropical forests, whereas aquatic biomes ranged from a mean of US$491 per hectare per year for the open ocean and US$352,915 per hectare per year for coral reefs.

So then, how does the value of ecosystem services relate to the economic impact of an invasive species? The services provided by an ecosystem are a function of its state. As we saw in the prior chapter, invasive species can cause ecosystems to transition between states, which means the services those ecosystems provide may also change. Given that an ecosystem provides numerous services, it is of course unlikely that any one invasive species would result in the total economic loss of services from an ecosystem. Still, the impact of just one invasive species on the state of an ecosystem can result in a considerable economic loss even if for only a single ecosystem service.

To provide a concrete example here and given our newfound familiarity with the dynamics of shallow lakes, let's consider the impact of just one invasive species in one ecosystem and for one ecosystem service: water clarity. Lake Mendota is a 40-square-kilometre lake near Madison, the capital of Wisconsin in the United States. The lake often exhibited poor water clarity due to nutrient run-off from local farming—as you will recall, nutrient run-off stimulates algae growth. However, *Daphnia pulicaria*, a native water flea that feeds on algae, once existed at such abundances that Lake Mendota would experience a clear water phase in spring each year, which provided considerable recreational value to the local community. In 2009, the invasive spiny water flea (*Bythotrephes longimanus*) was discovered in the lake and, due to its predatory nature, it reduced the *Daphnia* population by ~60 per cent, which subsequently led to an increase in algae and a dramatic reduction in water clarity year-round. Using survey data, in 2016 Jake Walsh and colleagues estimated that losing the clear water phase in the lake was worth about US$140 million to the local community, approximately US$650 per household.

Because economic impacts of invasive species are measured and presented in dollars, the severity and magnitude of these impacts are perhaps the easiest of the three categories to appreciate. Presenting impacts in dollars also makes it rather obvious why they are considered undesirable. Several estimates account for the total financial loss due to invasive species including the direct losses such as reductions in agricultural yield, and the indirect losses in the form of altered ecosystem services. In the United States, this estimate stands between US$100 and US$200 billion annually; Mariannne Kettunen and colleagues placed the cost to Europe at €12.5 billion (~US$15 billion) annually in 2009, which is most likely higher now; and invasive species cost approximately AU$10 billion (~US$7.3 billion) annually to the Australian economy. In 2021, Christophe Diagne and colleagues found that between 1970 and 2017 the total reported cost of invasive species worldwide had amounted to US$1.288 trillion. They also noted that this figure strongly underestimates the true cost and that costs continue to grow at an increasing rate as more invasive species arise via the pathways we described above. These costs are effectively an invisible tax on society; could that be considered as anything but undesirable?

Human health and wellbeing

There are numerous ways in which invasive species affect human health and wellbeing. Impacts range from the common invasive plant ragweed (*Ambrosia artemisiifolia*) producing copious quantities of allergenic pollen, to giant hogweed causing contact dermatitis, through to the psychological distress experienced by growers who are under financial stress from losing crops to invasive insects, or the stings and subsequent hospitalizations caused by the invasive fire ant (*Solenopsis* spp.). Human health impacts can also occur from the invasive species simply being unknown in the novel habitat and thus people do not know to avoid contact.

The silver-cheeked toadfish (*Lagocephalus sceleratus*), for instance, is a potentially lethally poisonous species of the pufferfish family (Tetraodontidae). They are native to the Indo-Pacific Ocean, including the Red Sea, and were first recorded in the Mediterranean Sea off Turkey in 2003 after migrating across the Suez Canal. The toadfish's population subsequently increased, and soon thereafter the first instances of human tetrodotoxin poisoning in the Mediterranean occurred. In December 2008, 13 cases of tetrodotoxin poisoning were registered by the Israel Poison Information Centre and at least seven people had died in Lebanon after consuming the fish. A 2011 survey in Turkey found that most customers attending open air seafood stalls were unaware of the species, that it was poisonous, and could not identify it, despite it being sold at the market. Given this risk the Turkish Ministry of Food, Agriculture, and Livestock banned harvesting of the species in 2012 although the invasive population in the Mediterranean Sea persists.

Perhaps the most serious human health impact that invasive species can cause is through the introduction of infectious diseases and parasites. The World Health Organization lists infectious disease as one of the leading global causes of human mortality, and invasive species both increase the number of human diseases in a location and increase the probability of people becoming infected. Zoonotic diseases—diseases that can jump from non-human animals to humans—resulting from invasive species introductions are particularly problematic, as raccoons (*Procyon lotor*) and roundworms demonstrate.

Raccoons were introduced to Europe in the early 20th century for the fur trade and to increase hunting opportunities, and, like many other mammals purposefully introduced, eventually escaped captivity and founded a non-native and invasive population. What importers also inadvertently received when they imported raccoons was the endoparasitic roundworm *Baylisascariasis procyonis*, which is ubiquitous in raccoons. The parasite lives and reproduces within the intestinal tract of raccoons, and its eggs are excreted within raccoon faeces. After two to four weeks in the soil, the worm becomes infective and can re-enter raccoons or, for that matter, any grazing animal, and it ends up back in the intestine to start the cycle again. However, in non-raccoons, once the invasive parasite is ingested, it penetrates the gut wall and migrates into other tissues, particularly the brain and eyes. Although serious health effects of *Baylisascariasis* in humans are rare, reported cases have resulted in vision problems due to neuroretinitis (swelling of eye tissues) as well as meningoencephalitis and encephalitis (swelling of the brain and its tissues), which have resulted in neurological problems and death.

Ectoparasites, such as fleas, ticks, and mites, are essentially parasites that live externally rather than internally like endoparasites. They too are often co-introduced with invasive animals and serve as vectors for some infamous human diseases such as typhus, Lyme disease, and the bubonic plague. The rise of West Nile virus in North America is a well-documented example of invasive species spreading an infectious disease.

West Nile is a single-stranded RNA virus in the genus *Flavivirus*—the same genus as dengue virus, yellow fever virus, and Zika virus—and was first discovered in 1937 in Uganda. Since then, episodic outbreaks have occurred across much of tropical Africa, southern Asia, and northern Australia. In 1999, West Nile virus was discovered in New York City, most likely co-introduced with its primary mosquito vectors, *Culex pipiens* and *Culex quinquefasciatus*, which appear to have hitch-hiked a ride from Israel to New York in shipping cargo. Within four years the virus had traversed North America. Between 1999 and 2010, nearly 2 million people were infected, resulting in roughly 360,000 instances of illness and 1,308 deaths. The United States Centers for Disease Control estimated that by 2014 West Nile had cost the United States US$778 million in long-term medical expenditures and loss of life productivity. Presently, West Nile virus is the leading cause of mosquito-borne disease in the continental United States.

Human health and wellbeing impacts are often reported in economic terms. Corey Bradshaw and colleagues estimated that invasive insects cost approximately US$77 billion per year globally, where a minimum of US$70 billion were costs associated with goods and services and approximately US$7 billion related to human health. While a monetary value is useful for determining the magnitude of some invasive species impact types, it is important to remember that the dollar value alone does not capture the full dimension of human health and wellbeing impacts. For some human health and wellbeing impacts, it is impossible to assign a dollar value at all.

To look at how the human health impacts of a single invasive species can transcend economically tractable to economically intractable, let's revisit the emerald ash borer which, as its name suggests, affects ash trees (*Fraxinus* species). In United States forests there are nearly nine billion individual ash trees, some of which are used to make a variety of products from bows to baseball bats and tools to tables. After American elm trees died off due to the Dutch elm disease invasion in the 1930s, green ash (*Fraxinus pennsylvanica*) were planted extensively as an urban street tree. Beyond providing a material resource, and like other urban trees, ash provide significant ecosystem services to residents in the form of shade, reduced air pollution, storm water control, and carbon storage.

When ash borer populations are sufficiently high, they often eliminate nearly all ash trees and the services they provide in an area. In economic terms, the cost of replacing ash trees on city streets ranges between US$1–4 billion for a single township. Since ash forests provide scenic and recreational value to homeowners, the loss of ash reduces home value and desirability and has been linked with a 1–2 per cent increase in crime rates. Since urban forests decrease stress in nearby residents, encourage physical activity, and improve air quality, each of which correlates with cardiovascular and respiratory health, the invasive ash borer naturally reduces such benefits. This outcome is sadly borne out in mortality data, as Geoffrey Donovan and colleagues found. They examined 15 US states and determined that more than 21,000 deaths related to cardiovascular and lower-respiratory tract illnesses were associated with the loss of native ash forests in urban areas due to the emerald ash borer invasion.

Invasive species impacts can transcend even the life of an individual person as emerald ash borer also causes cultural impacts. Many Native American and First Nation peoples have long traditional associations with black ash (*Fraxinus nigra*). The Wabanaki tribes in Maine and the Saint Regis Mohawk Tribe in New York and Ontario, Canada, for example, have deep cultural ties to black ash since it is used for basketry and for spiritual teaching. Although ash borers have not yet spread to most forests used by these tribes, the few locations that have experienced infestations report high losses of black ash with very little, if any, regrowth of trees after the ash borer's arrival. The loss of black ash makes it more difficult for tribes with an association to these trees to continue their cultural traditions, whether that be in the creation of cultural artefacts or the transmission of stories. The magnitude and meaning of this impact can hardly be captured in economic terms it would seem, and perhaps we shouldn't try.

Environmental

Any invasive species impact that is not directly economic or human health related is an environmental impact. Reductions in native species' populations, the loss of species entirely, or the shift in ecosystem states are the primary examples here. In 2016, Tim Doherty and colleagues examined the impact of invasive mammalian predators and found they were implicated in the extinction of 10 reptile, 45 mammal, and 87 bird species since

1500 CE. That's 58 per cent of the extinctions that have occurred in these vertebrate groups globally since then. In the same year, Céline Bellard and colleagues found that invasive species were identified as a key threat in a quarter of all threatened and endangered amphibian and bird species, 18 per cent of reptiles, and 15 per cent of mammals. It is likely, therefore, that invasive species will be implicated in future extinctions.

These examples are clearly environmental impacts, but are they undesirable? When impacts affect us directly, concluding that they are undesirable seems rather self-evident. Arguing that the loss of agricultural yield or the billions spent in treating and preventing human diseases associated with invasive species are not undesirable is hardly tenable. This is because many (especially in Western cultures) take the anthropocentric position for granted: the human-centred view where people and impacts to them are of primary concern. But ecological interactions and extinction are natural processes that have occurred ever since species existed. Because of this, the undesirability of environmental impacts may not be so obvious.

Let's consider a specific example here, the extinction of the robust white-eye (*Zosterops strenuus*). It was a green bird of about 7 to 8 centimetres in length that was endemic to Lord Howe Island located off the coast of Australia. The robust white-eye was common until in 1918 a ship ran aground, which inadvertently introduced the black rat (*Rattus rattus*) to the island, and by the mid-1920s the robust white-eye was extinct. The bird arguably provided no economic or obvious human health value to the few people who lived on Lord Howe, and it was not an aesthetically unique bird, being one of more than 100 species of white-eye in the genus *Zosterops* that are distributed throughout tropical Africa, south-eastern Asia, and Australia. Was that an undesirable impact? Of course the aviphile would say as much, and our moral intuition would likely have us agree, but on what grounds can we say that the presence on Lord Howe Island of the robust white-eye and absence of black rats is preferable to the absence of the former and presence of the latter?

Articulating answers to such questions is the domain of environmental ethics. Although it was built on the foundations of moral philosophy and axiology, both of which have developed over the past few thousand years, environmental ethics is a relatively young discipline, having developed alongside the environmental movement of the 20th century. If we were to summarize the focus of the field in a single question, it would be: What, if any, are the duties and obligations of moral agents, specifically humans, to the non-human domain? Attempting to give even an overview of the field is beyond the scope of this book. Instead, we will explore several arguments and their philosophical underpinnings that are often used to support the claim that environmental impacts caused by invasive species are undesirable.

When claiming that an impact is undesirable, we are making an implicit claim about value. As such, the most fundamental and important distinction separating the numerous positions in environmental ethics is in the type of value that non-human entities possess. By type of value, we mean whether something is valued 'instrumentally' or 'non-instrumentally' (hereafter intrinsically). If something is mainly valued instrumentally, then we are saying that its value is derived from the utility or benefits it provides us, often phrased as it provides a *means to an end.* If something is said to mainly have intrinsic value, then we are saying its value is independent of any instrumental value it provides us and it is valued in and of itself, often phrased as being an *end in itself.* A particular object might be valued both intrinsically and instrumentally, but the type of value ascribed or seen as most important depends upon an individual's ethical position. Additionally, even though two people may draw the same conclusion about the undesirability of invasive species impacts, the reasons for those conclusions may be different and have different implications.

Traditionally, the moral obligations of humans towards the non-human environment were based on anthropocentric paradigms where intrinsic value was attributed to humans alone, or at a very much greater level than ascribed to non-humans, and non-humans were primarily valued instrumentally. It is this type of value that underpins the undesirability of the economic and most human health impacts of invasive species that we saw above. Yet, this duality creates an obvious problem for the instrumentalist when considering the undesirability of environmental impacts. On what grounds can the environmental impacts of an invasive species be considered undesirable when there are no obvious human-related impacts? As it happens, numerous arguments based on instrumental value do conclude that environmental impacts of invasive species are undesirable. One line of reasoning that has garnered widespread support both publicly and politically comes from *enlightened anthropocentrism*, which claims that our duties towards the environment derive from our duties towards other humans. So, in the specific case of invasive species, even where no tangible economic or human health related impacts are evident, environmental impacts of an invasive species are nevertheless undesirable because human society and wellbeing is dependent upon functioning ecosystems and a sustainable environment.

Not everyone finds the instrumentalist position compelling for environmental protection. Many environmental ethicists see it as weak or inconsistent, and seek to resolve those philosophical issues by developing arguments based on intrinsic value. It is commonly agreed that moral agents are obliged to protect objects possessing intrinsic value. As such, once intrinsic value has been ascribed to the non-human environment, then concluding that the environmental impacts of invasive species are undesirable naturally follows. However, while it is generally taken for granted that humans possess intrinsic value, extending that to non-humans has its difficulties.

A key conflict in ascribing intrinsic value to non-humans is to what biological level should intrinsic value be ascribed. When considering humans, intrinsic value is possessed by the individual. But in environmental ethics, the dominant paradigm is ecocentrism, where ecosystems or populations of species are ascribed intrinsic value of greater import than the intrinsic value of their component parts, the individual organisms. In the invasive species context, ecocentrists simply argue that given our knowledge of ecosystems, the presence of an invasive species is almost certainly affecting other populations or the ecosystem itself, and therefore impacts on those entities with intrinsic value are undesirable.

Upon reading that you might say: if all populations have intrinsic value, then concluding that an invasive population impacting upon a native population is undesirable requires that the native species has greater value than the non-native. And you would be right. The 'natural-historical' view of intrinsic value maintains that non-human entities have intrinsic value by virtue of their independence from human interference and design. That is, *naturalness* is value-adding. For invasive species that depend upon human transport pathways to reach novel habitats, the non-native species' population does not exhibit that naturalness, that value-adding trait. Similarly, ecosystems that are in a more natural state are valued more than those modified by humans. In fact, even if we have two ecosystems that are currently identical in every way, but one was rehabilitated while the other has gone untouched, the rehabilitated ecosystem will have less value than the historically untouched ecosystem according to the natural-historical position.

Some theorists reject ecocentrism for biocentrism, whereby intrinsic value is possessed by individual organisms and not collectives. The problem with ecocentrism, a biocentrist might claim, is the value of an ecosystem or population is a by-product of the value possessed by the individuals of non-human species, and, secondly, it is logically inconsistent. If we value humans individually, why should we only value chimpanzees as a collective and not as individuals? In the context of invasive species management, this subtle shift from the collective to the individual can result in irreconcilable differences between these positions; especially where lethal control of invasive species to limit their impact is concerned. An invasive species having some observable impact on an ecosystem, to the ecocentrist, must be removed given the intrinsic nature of the ecosystem. To the biocentrist, although they acknowledge the environmental impact of the invasive species, killing individuals of the invasive species to reduce that impact is generally not acceptable and alternative management solutions should be sought.

Let's now return to the robust white-eye on Lord Howe Island and why rats causing their extinction might be considered undesirable. The enlightened anthropocentrist, who bases their arguments on instrumental value, might argue that the loss of the white-eye represented a loss of experiences available to the human species, and the introduction of the black rat was no substitute. The ecocentrist would most likely argue that the robust white-eye, especially as an endemic species, possessed greater intrinsic value than the black rat, which is common in many places, and thus its loss makes the world poorer. A biocentrist would likely agree with the ecocentrist, although they might not condone the lethal management action required to eradicate the rats.

We started this book by defining an invasive species as a non-native population that either has spread beyond its initial introduction location or that causes undesirable impacts. At the time, we could only point to social values as being the reason such impacts were considered undesirable. Now, over the prior two chapters, we have seen how invasive species cause those impacts, and, in this chapter, explored the philosophical principles for why undesirable impacts are undesirable, whether they be economic, human health related, or environmental. Although the data are clear—invasive species can have significant negative impacts on other species and ecosystems—not all readers will agree on the importance of, or the threat posed by, invasive species because of differences in their philosophical positions. Still, most people, from the public to the political, find the arguments for the undesirability of invasive species impacts sufficiently compelling to warrant some sort of action.

Chapter 9
An ounce of prevention is worth a pound of cure

Some years ago, I (DW) travelled from Sydney, Australia to the island nation of Vanuatu, 2,500 km away in the South Pacific. I woke early, packed a snack for the four-hour flight, and then, as many people have before and after, promptly forgot about it. On descent into Vanuatu the flight crew handed me the immigration card that I sped through, ticking that I was not carrying prohibited goods; I had forgotten about the two bananas in my backpack. Even the very large signs displaying pieces of fruit with a red circle and slash across them did not jolt my memory. I got through to immigration and quarantine, handed over my documents, and walked out into the sunshine. In this completely innocent and absent-minded moment I had inadvertently defeated the first line of defence in managing invasive species, and simultaneously put Vanuatuan agriculture and their environment at risk.

Managing invasive species, or species that can potentially become invasive, is a logical reaction to their undesirable impacts. The management strategies employed by most governments roughly follow the invasion process that we have explored throughout the book—transport and introduction, establishment, and spread—and are referred to as prevention and eradication, containment, and resource protection (Figure 13). Notably, the cost of management action increases as the population size of the

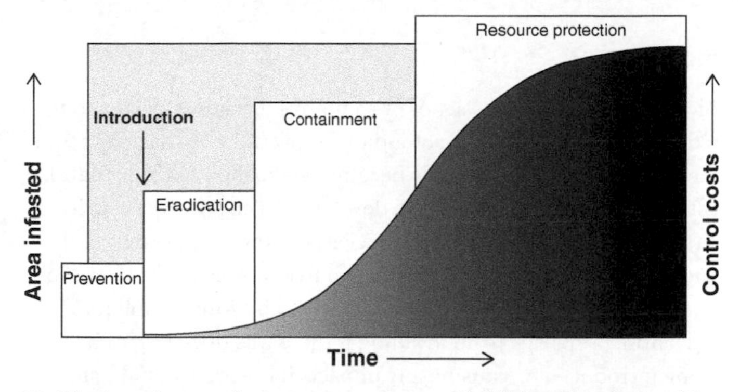

13. The invasive species management curve highlights the relationship between the area occupied by a non-native species, the costs associated with its control, and the change in management options available.

invasive species increases and becomes more widespread. Given that management resources are finite, this leads governments to balance the amount of effort and resources being applied to particular management strategies.

Before a non-native species is present in a novel habitat, the primary and least costly management option is preventing its arrival. This outcome is the aim of government biosecurity programmes and is only possible prior to the founding of a nascent non-native population. Once a population has established, assuming the population is small and occupies little area, it can be possible to eradicate the nascent population, which is where early detection and rapid response programmes are used. However, once the non-native population begins to spread beyond the point of initial introduction, the management goal switches away from eradication to containing the spread. If containment fails, or the species has spread sufficiently far by the time they are detected, management must identify key assets impacted by the invasive species and invest in their protection. Following is a detailed exploration of these strategies.

Biosecurity

Biosecurity refers to the set of preventative measures designed to reduce the likelihood that potentially impactful non-native species are introduced and ultimately become established within a nation, island, or state. The first step in developing a cogent biosecurity strategy is to determine which species pose the most threat and apportion the finite management resources accordingly. It would not make financial sense, after all, to spend billions of dollars preventing a species from invading if there was little risk of it being introduced or causing any impacts if it were to establish.

Determining the threat that a species poses is carried out using risk assessment tools. A risk assessment accounts for the likelihood that an event will happen and the probable severity of its impact. For invasive species, that means determining the likelihood that a species would be introduced through a particular introduction pathway, the probability that it would become established, and an assessment of the potential damage that it might cause. High-risk species are those that have a high likelihood of being introduced and establishing a non-native population, which would then cause considerable impact. The risk assessment also identifies the high- and low-risk introduction pathways for assessed species. This effort allows biosecurity agencies to appropriately target pathways that are most likely to carry high-risk species. We know, for instance, that the highest-risk pathway for introducing non-native insects is through the import of fresh fruit and vegetables for food markets and not the pet trade.

Risk assessments have two principal applications in management. First, they provide guidance on which species are, or should be, permitted to enter a jurisdiction, and the most likely pathways by which a species would enter. Some countries use a blacklist approach and ban only those species that present the highest risk, an innocent until proven guilty approach. Other countries, Australia and New Zealand for instance, use a white-list approach and prohibit all species from entry except those with demonstrable low risk. Second, risk assessments are used to guide policy in latter stages of management and coordinate with other nations to reduce invasion risk collaboratively. Risk assessments require assembling broad evidence on a species: its suitability to various climates, whether it is invasive elsewhere, the type and magnitude of impacts it causes, and successful control measures, all of which are useful information if a species were to become invasive.

One example of 'collaborative pre-border' risk reduction is seen in the treating of wood-based materials used in packaging, pallets, crates, or dunnage when transporting products between countries. These materials can harbour wood-boring insects that can affect numerous industries and the environment. Through the International Plant Protection Convention of the World Trade Organization, member nations require these packaging materials to be heat treated, or a suitable pesticide applied, to kill stowaway insects. Enforcement is enacted through a certification process and member nations can deny the entry of any wood packaging into their jurisdiction if these certifications are not provided.

International collaborations for handling ballast water provide another example. Cargo and transport ships use wet ballast to maintain stability of the vessel while under way and, as we noted earlier, this has led to the introduction of numerous species to ports all over the world. To reduce the risk of invasion from this pathway, the International Maritime Organization mandates that international ships practise open-ocean ballast water exchange. This action requires ships to replace ballast water that may have been taken on in an estuary, where the ship loaded or unloaded goods, with water from the open ocean. The biology underpinning this policy recognizes that most aquatic organisms have a tight range of water salinities and temperatures in which they can survive. Species that tolerate the relatively warm and less-saline waters of an estuary typically cannot survive the more saline and cold waters typical of the open ocean, and vice versa. By strategically locating where ballast water is exchanged, the risk of species within ballast tanks establishing within the estuaries of ports is much reduced. Evidence thus far suggests that these regulations have reduced the rate at which non-native species are establishing around ports due to ship ballast water.

Such policies and strategies are not flawless. Biosecurity may not inspect someone's luggage, or companies and individuals may knowingly attempt to thwart biosecurity efforts by falsifying certifications to meet international regulation or by not declaring the contents of packages. Additionally, some options that reduce invasion risk in one situation are not permissible in another. Treating packaging with pesticides, for instance, may reduce invasion risk from potential stowaway insects, but the pesticide may have human health implications and therefore cannot be used on certain products. Continuing to find alternative suitable solutions is therefore an ongoing challenge. Still, given the sheer number of goods transported globally, reductions in invasion risk from biosecurity and preventative measures alone will never be zero, making other measures necessary.

Early detection and rapid response

When non-native species do make it through the biosecurity net, early detection and rapid response strategies seek to find and eradicate them before their populations become too large. Quick action must be taken since the difficulty and cost associated with eradication rapidly increases as the non-native species' population grows and spreads to more areas (Figure 13). Given society's track record with causing the extinction and extirpation of numerous species, it would seem that eradicating a single species would be simple. But this is not necessarily true, especially since the cure cannot be worse than the problem. Clear-felling a forest to remove an invasive plant is not a viable strategy, as an obvious example. Consequently, not all eradication efforts are successful. Since the 1890s, of the 672 eradication attempts targeting 130 invasive arthropod species, 395 were declared successful while 110 failed; the remaining attempts are still in progress (Figure 14). There is little data for other taxonomic groups, but they likely show a similar trend with some not insignificant portion of eradication attempts failing.

There are numerous reasons why eradication attempts fail where others succeed: the management agency may lack the necessary funds to put in sufficient effort, or they may not have appropriate equipment or personnel to carry out the biological surveys.

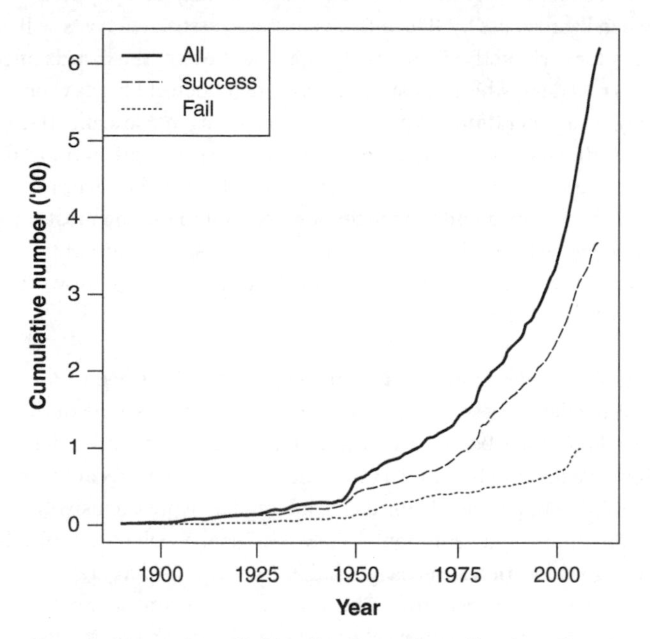

14. Cumulative number of initiated arthropod eradication efforts has grown exponentially since the 1950s. Also depicted is the cumulative number of declared successes or failures of those eradication attempts.

All else being equal, however, the main reason that eradication attempts fail is due to the difficulty in detecting non-native species in the landscape while populations are low. Without detecting them in the first instance, management agencies do not know that they need to carry out eradication activities, and once it is known that a non-native species is present, ongoing detection is required to carry out its eradication.

Our ability to detect a particular species is principally determined by the survey methods available to detect it. For any given species, there are often numerous methods that could be used to detect a non-native population. Small terrestrial lizards, for instance, could be detected by surveyors simply looking for the lizard, or by using small traps, or even by using cameras that take pictures when lizards pass by. The effectiveness and associated costs will vary for each method such that traps might detect the lizards once every 10 days, whereas using cameras might detect lizards every day. As a general rule, when sensitive methods that are effective at detecting individuals of a target non-native species are employed, eradication attempts are eight times more likely to be successful. Although management agencies always want to use the most effective methods, in practice some methods simply cannot be used due to resource limitations, or the technique may not work in a particular environment.

Despite these limitations and challenges, there have been spectacular successes in eradicating invasive species. One of the best known has been the eradication of invasive rats and other invasive mammals from islands, and the subsequent recovery of seabird populations. Seabirds are adapted to living exclusively within marine environments. They nest in large colonies, typically in rugged locations on coastal and island habitats that are naturally free of predators. This predator-free status of most seabird breeding locations changed over the last few centuries when the Polynesian rat (*Rattus exulans*), black rat (*Rattus rattus*), and brown rat (*Rattus norvegicus*) were introduced to about 90 per cent of seabird breeding sites globally. Since many seabirds nest on the ground, and some in burrows beneath it, these invasive rats devastated seabird colonies by consuming bird eggs, juveniles, and even adults. Of the 365 seabird species worldwide, more than 100 are at risk of extinction, and some families greatly so, with invasive predators contributing substantially to their plight.

In response, conservation scientists engaged in a worldwide effort to eradicate invasive rats and other non-native mammals from islands. The results have been remarkable. On Tromelin Island in the western Indian Ocean for instance, the red-footed booby (*Sula sula*) and the masked booby (*Sula dactylatra*), which were present before rat eradication, saw an increase of about 23 per cent in breeding pair numbers after rat eradication. Additionally, the white tern (*Gygis alba*), which hadn't bred on the island since 1856, and the brown booby (*Sula leocogaster*), which had never been observed nesting on the island, were recorded breeding there post-rat eradication. More generally, of the 251 invasive mammal eradications on 181 islands, 236 native species have benefited, including four species that had their risk of extinction status reduced, and a number of seabird colonies are now recovering their former numbers.

Stop the spread

When invasive species become more widespread management and policy actions naturally move towards slowing and ideally halting the spread of the invasive species. These efforts have become colloquially termed 'stop the spread' campaigns, and include continued eradication efforts to suppress population growth, and public outreach. The goals of public outreach are to provide citizens with a simple set of protocols that, if followed, reduce the risk that they contribute to the invasive species' geographical expansion. These campaigns often include quarantine zones that prohibit industry and the public's activities that might move individuals of the invasive species from where it is currently established, within the quarantine zone, to locations where it is not yet established, outside the quarantine zone.

Public outreach is a major component of slowing the spread since our everyday actions contribute to expanding the invasive species' range once it has established. Take recreational boating, for instance. When a boat is left in the water, at a marina or mooring, it will invariably accumulate a hull fouling community—organisms that take up residence on the underside of the vessel. Even though boat operators take measures to limit the growth of these organisms, they are never entirely absent. One study in a marina in British Colombia, Canada, found nine invasive species on the hulls of recreational vessels, including on vessels used regularly and considered otherwise 'clean'. When these boats are used, they inadvertently become taxis for those invasive species to reach new habitats. Even recreational vessels that are removed from the water and trailered from one location to another can be vectors. Water in the bilge, bait wells, or the engine provide refuge for small invasive species for short trips. The spiny water flea (*Bythotrephes longimanus*), the culprit invasive species that imposed financial costs on local communities in Wisconsin, appears to have expanded its range rapidly beyond the Great Lakes in North America with help from recreational boaters.

An important technique for combating these relatively short-range dispersal events, which cause satellite populations and rapidly increase the spread of non-native species, is delimiting surveys (Figure 15). Delimiting surveys start by sampling the landscape at a somewhat coarse geographic resolution; that is, survey locations are not tightly packed. When the invasive species is found, sampling is then focused in the area around the detected individuals to find the 'edges' of the invasive species' distribution. By finding the edges of the population, invasive species managers can then deploy eradication and treatment efforts much more effectively, which will have the greatest impact on the invasive

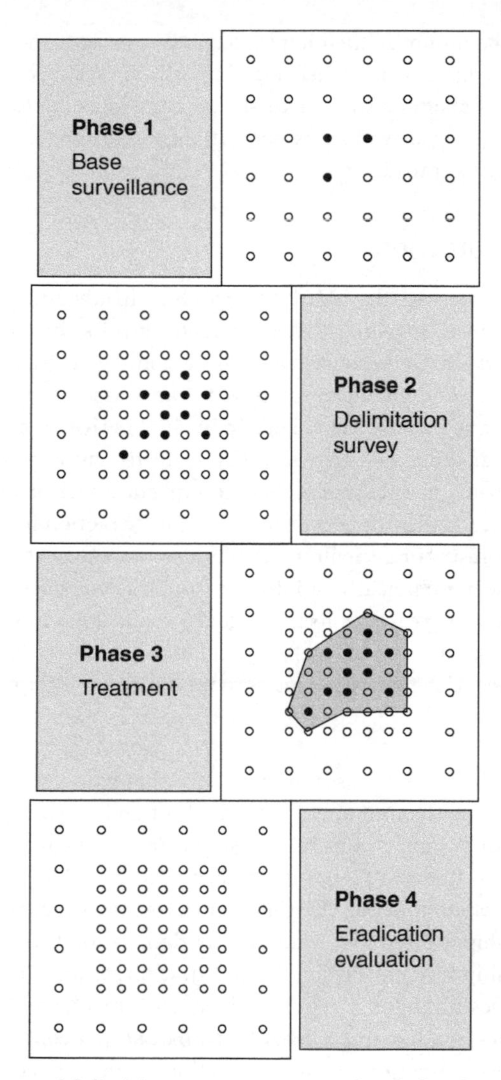

15. Conceptual delimiting survey strategy used to find and suppress satellite populations of an invasive species in a cost-effective manner.

species and minimize their impact on native species. Once treatments have been applied, high intensity surveying continues to ensure that efforts were successful at removing, or at least reducing, the invasive species' population, and ultimately stopping or slowing its spread.

Asset protection

In many instances, the invasive species is so firmly established and has spread so widely that attempts to slow its advancement can overwhelm the resources of management agencies. In these cases, policy and management move towards asset protection. Assets in this context can include individual native species that might be endangered, as we saw with seabirds, entire ecosystems that are being maintained in their natural state such as the Florida Everglades, or ecosystems used for extraction industries such as forestry or agriculture. It may also include protection of assets like human health or infrastructure that the invasive species may negatively affect. The goal of such programmes is not eradication, even of satellite populations, but to reduce the abundance of the invasive species to minimize their impacts on the asset.

A major challenge for asset protection is that while we know that reducing the number of individuals of an invasive species reduces their impact, this relationship is rarely linear, as depicted by line c in Figure 16. Instead, evidence that informs the 'impact–abundance curve' in Figure 16 shows that this relationship is often highly non-linear. Sometimes this relationship is in our favour, like line d, where a small decrease in a species' abundance rapidly reduces its impact, and sometimes not, like line a, where a drop of 80 per cent in abundance is required before a reduction in impact occurs. Most empirical evidence suggests that the same invasive species can show different relationships depending on the impact itself and in what ecological context it is measured.

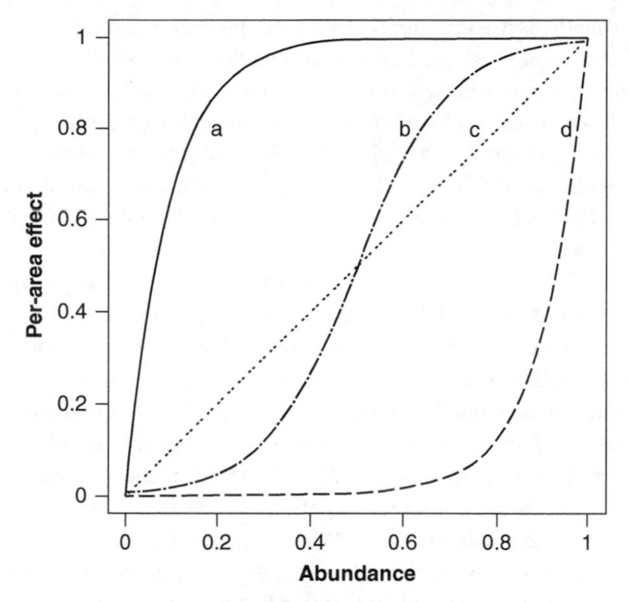

16. The possible relationships between the abundance (where 1 is the current abundance and 0 is eradication) and the ecological or socio-economic impacts (measured on a per area scale, where 1 is maximum impact and 0 is no impact).

Research on the impact-abundance curve for a variety of invasive species and locations has only just begun, and so it is not clear if one type of relationship is more common than another. Most asset protection efforts simply assume that fewer individuals of an invasive species are better. Although true in principle, it does not necessarily lead to an efficient use of resources. The only prioritization in effort that typically occurs is by ranking the relative value of the affected assets and spending on those that garner public interest. There is a lot of room for research on the relationship between abundance and impact of invasive species, with the practical outcome of this research being adoption of evidence-informed decisions by managers on how best to prioritize and carry out asset protection efforts.

The methods used to suppress invasive species around valued assets mirror those used in eradication efforts—physically removing organisms, altering the habitat to make it less suitable, and using chemical or other treatments to kill the organism *in situ*—and include the deliberate release of 'enemy' species. Biological control involves identifying an ecological enemy of the invasive species, often a non-native species itself, and releasing it in the non-native range where the invasive species is well established. Historically, due to inadequate vetting of the enemy species, this approach has resulted in some spectacularly detrimental invasive species: cane toads (*Rhinella marinus*) in Australia, the Javan mongoose (*Herpestes javanicus*) in Hawaii, and the harlequin ladybird (*Harmonia axyridis*) in Europe are some of the more infamous cases. Modern biological control efforts, however, are heavily regulated and must meet stringent requirements to ensure that only the target invasive species is affected before being released. When done right, such programmes have been greatly successful. On average biocontrol agents reduce invasive plant biomass by 82 per cent and reduce invasive insect abundance on crops by 130 per cent as compared with control groups.

In a similar vein to stopping the spread, asset protection options include citizen science or citizen participation programmes to help detect and suppress invasive species. Some programmes, such as those in Florida to combat the lionfish (*Pterois volitans*) invasion or those in Australia to reduce cane toad (*Rhinella marinus*) numbers, include competitive derbies whereby participants compete to capture the most or the largest individual(s) of an invasive species to win prizes (Figure 17). These events serve the dual purpose of affording organizers the chance to elevate the importance of invasive species management and policy to a general audience, while also reducing the abundance of key invasive species. To what extent they protect valued assets is an open question, and the impact they have on the invasive species' population may be marginal. Still, if nothing else, they raise

17. Advertisement for the lionfish derby run by the Reef Environmental Education Foundation (REEF.org) held at cities along the coast of Florida, USA. Lionfish derbies inform citizens on invasive species generally and lionfish specifically. Organizers coordinate and encourage targeted removal of lionfish by citizens by offering prizes for lionfish catch on the event day. Removal of lionfish reduces the undesirable impacts they have on coral reefs.

awareness of the invasive species issue that in the longer run may reduce the likelihood that new non-native species are introduced.

As for the accidentally smuggled-in fruit, it was the following morning by the time I (DJW) became aware of my error. I put my hand into my backpack to fish out fresh clothing, only to return with a fistful of brown, mushy banana. And, fortunately, that was all. There were no stowaways in the mess.

Chapter 10
It is never simple

In laying out how species become invasive, cause undesirable impacts, and how they are managed, we have had to gloss over many complexities along the way. Exploring each complexity as it appeared would have resulted in a book three times as long and the main concepts would have been hidden in a quagmire of caveats and hedging statements. While the main concepts discussed hold in most cases, there are almost always exceptions or competing values and goals that make the identification and management of invasive species complex. Here, we turn our attention to several of these complexities, leaving the reader to ponder potential solutions, though we warn that there are currently few.

Weighing the desirability of impacts

If some non-native species cause undesirable impacts, then it stands to reason that some non-native species must not cause undesirable impacts, or perhaps even cause desirable ones. This is certainly true. In the North American Great Lakes, the zebra mussel has improved water clarity and become a food source for some recreationally valuable native fishes. For recreational divers who value clear water, since it makes underwater observations more enjoyable, for recreational fishers attempting to catch those native fishes feeding on zebra mussel, and for the businesses that support these activities, these are desirable outcomes of the zebra mussel invasion. And yet, as you will recall, we spent several paragraphs earlier in the book highlighting the undesirable impacts of the zebra mussel invasion in the Great Lakes.

Herein lies the complication: a non-native species in a single region can cause both desirable and undesirable impacts, especially after it has been established in a location for decades or more, as have zebra mussels. Sticking with just marine invasions, Stelios Katsanevakis and colleagues examined the impacts to ecosystem services caused by 87 non-native species found in European seas. Across nearly every category of service, from food to aesthetic value, both undesirable and desirable impacts were observed (Figure 18). Ross Shackleton and colleagues conducted a similar analysis on the impact invasive species have on human livelihoods and wellbeing by examining 37 invasive species that are established globally, primarily plants, and found that 48 per cent had both undesirable and desirable effects, with six species primarily having desirable ones.

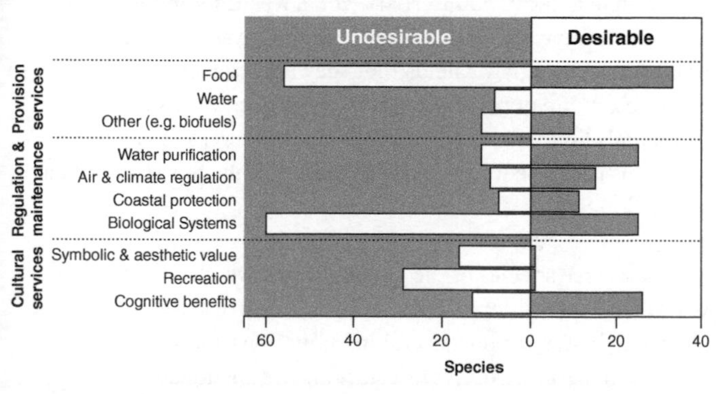

18. The number of non-native marine species in European seas that cause undesirable and desirable impacts on ecosystem services. Some non-native species may be counted multiple times, since they can produce both desirable and undesirable impacts across multiple ecosystem services.

How do we manage these non-native species? If the zebra mussel improves water clarity, which is good for recreational divers, but at the same time reduces beach access, since the mussel encrusts surfaces below the high-water line, then removing the mussel favours beach goers, while not removing them—itself a viable management action—favours the interests of divers. Such a simple trade-off is rarely the case since many aspects of the natural or human environment are affected by invasive species simultaneously. One path forward is to aggregate all impacts and determine the overall cost or benefit to a community, a strategy not without its own problems or biases. How do we weigh the impacts felt by different parts of society, say a wealthy or poor community, for instance?

People living near or below the poverty line may be heavily reliant on locally available trees for their wood, fuel, and fibre needs. Whether these needs are satisfied by an invasive or native tree does not concern them. In fact, an invasive species may be preferable for their needs as it may grow quicker and be easier to harvest than a native counterpart. Yet, a wealthier individual living nearby may view the presence of the invasive tree as a problem as it overgrows fields that they use for recreation, reduces native vertebrate populations that they aesthetically value, and compounds fire risks around homes due to increased fuel loads. While this might seem like a contrived example, this scenario has occurred in South Africa.

At least 23 tree species of the genus *Acacia*, which are native to Australia, are considered invasive in South Africa. Acacias have had a long history of purposeful introductions and were originally used within agroforestry, reforestation, and sustainable development initiatives in the late 1800s. Depending on the species, the value of acacia trees is in their production of wood and pulpwood, and use of their tannins for leather-making, perfumes, gum, and food. They are still commonly sold as ornamental plants and used to stabilize soil within ecosystems that are deemed in need of reforestation or restoration.

After introduction, acacia trees easily spread as they are highly flexible in their ecological requirements. In many locations the trees reach high densities, crowding out other trees and understorey plants, and turning habitats into near monocultures. This act clearly has negative impacts on native plant communities and contributes to both increased fire frequency and altered water flow. These local and regional changes cascade through ecosystems to produce a variety of undesirable impacts to the human communities living within and around acacia forests, some of which include decreased local water supply, a reduction of flora and fauna diversity, decreased fibre and food resource diversity, and a reduction in tourist experiences and tourism generally.

But not all segments of society have the same experience. In less wealthy regions of South Africa, silver wattle (*Acacia dealbata*) is used for a variety of needs: as a ready source of firewood for heating and cooking, to build and repair fences that contain livestock, in the creation of tools and other household furnishings, for use in medicines, and as domestic animal fodder. Although the people using silver wattle recognize that it can grow in locations where it is problematic and lead to problems that even directly affect them, there is no question as to the benefits that this tree provides these communities. As such, simply removing acacia is not a suitable solution. Without acacias, the community's fuel and fibre needs would have to be met by another source, and given there is no suitable alternative native species, then another invasive species would be required. What should invasive species managers do? Or, perhaps framing this as an invasive species issue is too narrow?

Invasive species and endangered species

As we have pointed out throughout this book, invasive species can negatively affect native species by consuming them, competing with them for resources, or by altering habitats and making them less favourable to native species. The urgency to deal with invasive species in these situations increases when the native species or the ecosystems being affected are endangered. Removing the invasive species in such cases becomes an obvious and necessary management strategy if we wish to conserve those threatened entities. But, what if the invasive species, despite generally causing negative impacts, happens to affect an endangered native species positively? Here, there is a conflict between limiting the impacts from the invasive species, while at the same time maximizing the recovery of the threatened species.

An example of such a conflict occurred in San Francisco Bay between management of an invasive saltmarsh cordgrass, of the genus *Spartina*, and restoration of the native and endangered California clapper rail (*Rallus longirostris obsoletus*). In the 1970s, the United States Army Corp of Engineers, in an attempt to reclaim lost marshland, introduced to San Francisco Bay *Spartina alterniflora*, a cordgrass native to the Atlantic coast of the Americas. The introduced *Spartina*, however, outcompeted and hybridized with the native cordgrass and invaded about 320 hectares of the bay. The invasion altered habitat characteristics, affected food web dynamics, and negatively affected a multitude of native species and ecosystem functions.

The California clapper rail, however, which lives in marshland habitat, specifically in mature stands of cordgrass, is one species that benefits from the invasive *Spartina*. The rail became endangered when urban expansion removed large tracts of its native habitat, stands of native *Spartina*. Although invasive *Spartina* does not perfectly replicate the rail's original environment, it is sufficiently alike to provide the rail with suitable habitat.

To restore the ecosystem to its original state, invasive species managers would need to remove the invasive *Spartina* and its hybrid and allow native *Spartina* to recolonize those habitats. But this is a process that takes time, on the order of years. And since the rail requires mature stands of either species of *Spartina*, there would be a considerable interval during the transition whereby the rail would be without suitable habitat, placing the rail at further risk. In this specific example a solution exists that allows wildlife managers to navigate the conflict—by carrying out removal and restoration activities slowly and in stages—but solutions are not always readily available in practice, and what works for one species or one ecosystem may not work in another.

A potentially more complicated case can arise when an invasive species is endangered in its native distribution. Consider the wattle-necked softshell turtle (*Palea steindachneri*), endangered in its native home of south-eastern Asia due to overharvesting for food and traditional medicines, yet invasive on the Hawaiian island of Kauai, where it competes with and consumes native fishes and presents a potential threat to the endemic and already endangered Newcomb's snail (*Erinna newcombi*). Or *Juniperus bermudiana*, the Bermuda cedar, critically endangered on its native home of Bermuda due to the introduction of scale insects that attack the tree, but thriving on the islands of St Helena and Ascension in the southern Atlantic. Or the barbary sheep (*Ammotragus lervia*), which has been introduced to numerous locations globally, primarily for hunting, while its numbers decline in its native distribution of northern Africa.

In each of these, and other similar, cases conservation and invasive species researchers, regardless of the value positions they hold, face the ethically tenuous position of trading one species for another. From a conservation perspective, the invasive population provides the endangered species with an insurance policy against extinction; for instance, we can imagine individuals in the native range being wiped out by wildfire or flood, which without the invasive population would have meant the extinction of the species entirely. Perhaps then, provided the invader is not driving the imminent extinction of a native species, retaining it would be a preferable solution in the short term. But ecosystems are highly complex and our ability to predict future states is very limited. There are no easy answers.

The impact of climate change

Somewhere between 25 per cent and 85 per cent of the roughly nine million eukaryotic species on Earth are on the move, and climate change is the culprit. As the climate changes, becoming warmer in most places and cooler in a few, while rainfall patterns change from the historical norm to something rather new, the distribution of many species will naturally shift as they attempt to maintain suitable climatic conditions to survive. In the northern hemisphere, many species have been recorded moving northwards or seeking higher elevations as their historical native distribution has warmed. The changing climate will also remove biogeographic barriers. For instance, disappearing sea-ice in the Arctic has resulted in some marine mammals and birds making the journey between the Atlantic and Pacific Oceans. In the coming decades, this shift in species' distributions will present a major challenge for society since the change in species compositions and interspecific interactions within these habitats will ultimately affect ecosystem services and, by extension, human wellbeing and livelihoods.

Species shifting their ranges due to climate change also complicates how we determine which species are and are not invasive. We defined an invasive species as one that is not native to a region and that causes undesirable impacts; and a non-native species as one that humans have directly or indirectly facilitated to cross a biogeographic barrier to a region outside of its native distribution. If we follow this definition strictly, then since human activities have been a major driver of climate change, any species that manages to cross a biogeographic barrier due to climate change might be considered invasive. But, is this definition too blunt an instrument to determine which species are and are not invasive under climate change?

Let's consider the grey whale (*Eschrichtius robustus*). Its historical distribution was the northern Pacific Ocean, along the continental margin from Japan in the west to Mexico in the east, where it currently exists, and a sister population in the north Atlantic that went extinct due to whaling in the 18th century. Evidence from ancient DNA suggests that the two grey whale populations mixed several times over the past several million years during periods of reduced sea-ice coverage. Since the extinction of the Atlantic population, Arctic sea-ice has prevented the Atlantic from being colonized by Pacific grey whales, but this is changing as the Arctic warms. In 2010, a single grey whale was sighted off Israel, and in 2013 another was sighted off Namibia. Assuming for the moment that these trans-oceanic wanderings continue and Pacific grey whales become established in the Atlantic, and given that whales generally can have large impacts on ecosystems, should the Pacific grey whale be considered invasive in the Atlantic? Or, since humans caused the extinction of the Atlantic grey whales originally, might we consider this a success story? An answer one way or the other is not necessarily obvious, although to date species that expand their range due to climate change are not considered invasive. Climate change, and the concomitant shift in species distributions, will require greater nuance in determining what is and is not an invasive species.

These issues that complicate how we manage and define invasive species are but a few of those that occur in invasion and conservation research. The interface of human society and the environment, in all its forms, is inherently complicated. It is the nexus where different human values clash and where imperfect information about the processes of nature obscures ideal solutions. Fortunately, there is much effort going into the development of new tools that will provide better resolution of the problems and hopefully allow solutions to be found, and much debate between scholars, policy makers, and interested members of society about how best to resolve these 21st-century challenges.

Chapter 11
A deliberate future

The story of invasive species is one of the development and growth of human society over the past 500 years. As new lands were sought for trade and territory, organisms of every type accompanied us; we brought some for trade, some for sport, some as reminders of homelands left behind, and some came without our knowledge. The modern world now bears that legacy. Invasive species have wrought enormous economic and human health impacts and are one of the greatest threats to other species and ecosystems, damage that affects us too in both the short and long run. Surely, then, have we not learned our lesson? Will we saddle future generations with ever more invasive species, or will new technology or policy decisions relegate invasive species to a historical footnote? Proclamations on the future of invasive species are difficult since they are intrinsically related to the actions of human society; still, using current trends we can make a good guess at how the next chapters in the invasive species story might unfold.

Will there be any 'natural' left in nature?

The prediction we can be most sure of is that the future holds more, not fewer, invasive species, at least in the near to medium terms. There are several good reasons for thinking this: first, we are a global species. Even if we only considered non-native species that hitch-hike to new areas when goods and people are transported, then we are bound to see more invasive species. In their 2021 outlook, the International Transport Forum at the Organization for Economic Co-operation and Development (OECD) estimate that both passenger and freight transport will more than double from their 2015 levels by 2050. More shipping and transport mean non-native species have more opportunities to reach new areas and expand their distribution. Developing countries seem particularly destined to experience massive increases to their number of invasive species as they engage more fully as global trading partners, since they have fewer resources to assign to preventative measures.

The second reason why we expect to see more invasive species is economic. To meet market demands in the pet, aquarium, and ornamental plant industries hundreds of millions of non-native organisms are kept in captivity, bred, and transported annually, both within countries and internationally. And these are large markets. According to an industry report, the flower and ornamental plant business is worth approximately US$50 billion annually. There is no reason to think that these markets will disappear soon and, as we have seen, private keeping is a key introduction pathway for numerous non-native species (Figure 3).

Finally, the issue of climate change. The interplay between invasive species and climate change is extraordinarily complex in large part because altered climatic regimes influence every part of the invasion process. To the degree that ocean currents and weather influence transoceanic shipping or airline flights, climate change will alter the networks through which non-native species are transported and introduced. Shipping through the Arctic Ocean, for instance, has already increased due to a reduction in sea-ice, and with it so has the probability that non-native species will be introduced to more northern latitudes.

The most obvious outcome of altered climates is that the suitability of habitats will shift from their current state. This will make some habitats, particularly those at higher latitudes or elevations, more suitable to non-native species from lower latitudes or elevations, and thereby make those habitats more susceptible to non-native species. Some environments may see shifts in suitability such that the invasive species currently occupying them may reduce in number. This will provide us with opportunities to reduce or even eliminate long-established invasive species. Two of the five most damaging freshwater invasive plants in South Africa, for instance, are predicted to contract their geographical ranges in the coming decades in response to climate change. If local environmental managers can capitalize on this reduction through increased control and eradication measures, they may be able to substantially reduce the undesirable impacts caused by these species in coming years.

However, the selection process that sees some species become invasive while others do not may have already tarnished this silver lining. There is some concern that invasive species may, as a whole, be more resilient than their native counterparts to the effects of climate change. This concern arises from the fact that many invasive species have biological traits that allow them to thrive in the sometimes-harsh conditions associated with transport and being introduced to sub-optimal habitats, and these traits may be just what is required to survive and thrive in a future world where extreme climatic events are common. What may be seen then is that both native and non-native species struggle with changing climatic conditions, but those species with broader environmental tolerances, which is often the case for invasive species, will prevail. There are simply many more ways for a habitat to be occupied by a resilient non-native species than not.

New impacts or more of the same?

One of the easiest predictions to make about the future of invasive species' impacts is found in our recent history. Céline Bellard and colleagues examined those species known to have become extinct, or extinct in the wild, since 1500 CE. While multiple drivers were typically identified for each species, invasive species were the most common driver of vertebrate extinctions, and second most common driver overall when including plant extinctions. So, as invasive species continue to increase in number, we can expect that the number of native species affected by invasive species, perhaps to the point of extinction, will also most likely increase. This process will undoubtedly affect ecosystem services, as we have already seen, and so adverse economic and human health impacts will almost certainly follow. Even without impacts to native species, human health impacts are particularly concerning given that many invasive species can be vectors for a variety of diseases.

What is difficult to ascertain at present is whether the magnitude of future impacts will be similar to historical observations. The drivers of environmental change—invasive species, over-exploitation of resources, removal of habitat for agricultural or urban development, and pollution—may interact synergistically in non-linear ways to produce much more serious effects than have been seen historically. Then, when we throw a rapidly changing climate into the mix, we add a layer of uncertainty with no historical precedent. Climate change has precipitated large-scale changes to a variety of biophysical processes that include raising the average air surface temperatures, increasing shallow- and deeper-water ocean temperatures, acidifying oceanic and coastal waters, altering precipitation patterns, and increasing the likelihood of extreme events such as flood, fires, hurricanes, and droughts. How these drivers interact is immensely complicated, and their outcomes highly uncertain. Still, we can be fairly certain in saying the impacts will be widespread and considerable.

On the upside

Breakthroughs in policy, technology, or our understanding may alleviate—or potentially amplify, if we get them wrong—the number of invasive species and their associated impacts that the future holds. Starting with policy: the rules governing the import of goods and the keeping and disposing of non-native organisms have certainly changed since the laissez-faire attitudes of the 18th and 19th centuries. Throughout the 20th century governments at both local and national levels implemented, wound back, re-implemented, and tweaked endless policies designed to protect their constituencies against biological invasions, with some serious missteps. As evidence-led policy is becoming more robust and widespread, fewer blunders should occur, but they will not be entirely absent. Generating unbiased evidence to inform policy takes time and decisions may need to be made on tighter schedules.

One area of policy advancing in a favourable direction addresses the trans-geopolitical nature of invasive species, particularly across contiguous jurisdictions. Invasive organisms have little interest in our political boundaries; as such the effectiveness of policies implemented in one jurisdiction is affected by regulations in their contiguous neighbours. If Switzerland is trying to eradicate Japanese knotweed (*Fallopia japonica*), an invasive plant occurring throughout mainland Europe, while France and Italy leave it unmanaged, then, even if Switzerland's efforts are successful initially, re-invasion is likely to occur. Fortunately for Switzerland, in 2014 the European Parliament passed regulations to ensure cross-border cooperation between member nations to prevent and manage invasive species. But such cross-border policies are yet to apply everywhere. In the contiguous United States, on average only 17 per cent of invasive plant species being managed in one state are managed by their neighbour. A growing body of evidence demonstrates the efficiencies gained from cross-border cooperation, so it seems reasonable that continued strengthening of interstate policies will occur.

Technological breakthroughs hold great promise for preventing introductions of non-native species and eradicating those that have established or become invasive. The two areas we expect to see most advancement in are detection, which is fundamental to both prevention and eradication, and eradication. Survey methods used to detect non-native species, such as traps or simply looking for the target species, are not perfectly effective; even if a target organism is present when surveyed for it may go undetected. This is particularly pertinent when surveying for non-native species during the early stages of invasion when they are, compared with the latter stages, rare.

The two ways to improve the probability of detecting a non-native species are, first, to increase the survey effort, or, second, improve the survey method itself. Increasing the effort sounds the simplest; merely deploy more traps or have more people look for the organism or look for longer. In practice, however, this is not necessarily possible. Each unit of effort has some associated cost that, since resources are always a limiting factor, results in a maximum amount of effort that can be practically achieved with a given survey technique. This trade-off is why methods research is imperative for not just invasive species research and management, but science generally.

In recent decades, environmental DNA has grown to be a powerful biodiversity monitoring tool. Environmental DNA refers to DNA deposited by an organism into the environment during normal activity; for instance, when it defecates or sheds its skin. This DNA can be collected and sequenced to determine the species present in an environment. While initially restricted to soil and aquatic environments, reductions to DNA processing costs and advancements in collection methods more recently have seen the use of environmental DNA expand to a slew of terrestrial species. Rafael Valentin and colleagues recently demonstrated that the spotted lantern fly (*Lycorma delicatula*), an invasive insect causing serious damage to native and important crop plants in the USA, could be detected using environmental DNA methods, which were also more sensitive than traditional visual searches.

Perhaps the greatest promise for invasive species management and eradication comes from another genetic technology: gene drives. Humans have long altered the genetics of domesticated and cultivated species to modify the traits they express—and many now scarcely resemble their ancestral form. Genetic modification in wild populations, however, particularly for invasive species management where the aim is to reduce the population, is much more difficult. Unless a gene is under positive selection (i.e. is favoured by natural selection), under normal inheritance the proportion of a population carrying the modified gene typically reduces through time (Figure 19). To be effective for invasive species management then, genetically modified individuals need to be released in large numbers to increase the frequency of the deleterious gene in the population. Gene drive technology, however, ensures that the modified genes, even if deleterious, are passed on to all offspring and increase in frequency in the population in successive generations (Figure 19).

The potential value of gene drive technology for invasive species management, then, should be clear: a manager need only alter the sex ratio of an invasive species, so that most offspring are male, and within a few generations the population will collapse. But what should also be clear is the potential loss that gene drives

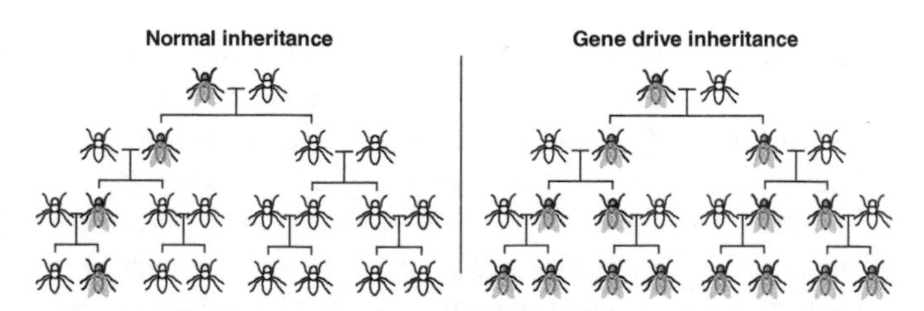

19. The proportion of flies carrying the modified gene (shaded flies) in a wild population typically reduces with time under normal conditions, unless it is being selected for. Gene drive technology ensures that the modified gene, even if it is deleterious to the population, increases in frequency.

could precipitate. Given that we are discussing invasive species, the glaring concern is that genetically modified individuals end up being introduced to the invasive species's native range, potentially resulting in its global extinction. Clearly, this would be unsatisfactory. The level of caution needed when employing these technologies *in situ* therefore cannot be overstated. We need to keep close to mind the well-meaning pronouncements once made of the promise of introducing non-native insectivores to control pest insects, only to have them become invasive species, such as the cane toad (*Rhinella marinus*) introduction in Australia.

Finally, what might the future hold regarding how we think about and understand invasive and non-native species? The most reliable prediction we can make here is on the future of invasion ecology as a discipline. As has been observed in numerous scientific endeavours, invasion ecology too seems destined to become more interdisciplinary—perhaps as a matter of necessity. Invasion ecology sits at the confluence of several streams of research: economics, ecology, evolutionary biology, public health, urban planning, and ethics, to name the dominant disciplines. Unpicking the confusing array of socio-environmental factors that knit together to complicate biological invasions, invasive species management, and conservation more broadly will require research teams that span research silos.

As our understanding of ecological systems advances, we may come to see non-native species in a very different light. If future environmental degradation outpaces the evolutionary adaptation of native species, some environments may effectively become uninhabitable to native species, leaving non-native species as the primary option for ecosystem restoration. It has happened already. In 2007, Aldabran giant tortoises (*Aldabrachelys gigantea*) and Madagascan radiated tortoises (*Astrochelys radiata*) were successfully introduced to a nature reserve in Mauritius to help control invasive plants and disperse the seeds of native plants. These functions were once performed by Mauritian giant tortoises (*Cylindraspis* sp.), which went extinct in the 19th century, and returning these processes to the ecosystem has improved ecosystem health. In the extreme, entire ecosystems might be constructed from species with disparate evolutionary histories to ensure ecosystem services are conserved.

Welcome to the Anthropocene

For almost all of life's roughly four-billion-year history on Earth, the distributions of its species were largely dictated by abiotic pressures such as climate and plate tectonics, coupled with biotic interactions such as competition and predation. Biogeographic barriers grew and fell on time scales of tens of thousands, if not tens of millions, of years. And species evolved and went extinct on similar scales as they competed with others in their ecological neighbourhoods. Then, in the briefest of moments, a mere 500 years, one species, *Homo sapiens*, single-handedly altered the trajectory of Earth's biogeographic and evolutionary future. We reshuffled the biological deck and increased the prevalence of a few dozen species globally at the expense of many.

Biotic homogenization, however, is but one aspect of the profound influence that *Homo sapiens* has had on the Earth. Since 1750, humans have released 555 billion metric tonnes of carbon to the atmosphere and thereby increased carbon dioxide to levels not seen for 800,000 years, and possibly much longer. The Haber–Bosch process, which converts atmospheric nitrogen to fertilizer, coupled with the burning of fossil fuels, has produced the largest impact on the Earth's nitrogen cycle since it originated 2.5 billion years ago, leading to widespread eutrophication in coastal and inland waters. Although the net global primary productivity has remained roughly constant, between 25 and 38 per cent of it is now appropriated by just one species: ours. It is not merely environmental agitprop to label the current epoch as the *Anthropocene*. Since life began on Earth, no single species has had such prodigious impact on all others.

Many claim that humans have become a *force of nature*. Oliver Morton, in *The Planet Remade*, argues that this leads to a paradox: 'humans are grown so powerful that they are a force of nature—and forces of nature are things which, by definition, are beyond the powers of humans to control'. This rather ominous and pessimistic characterization is not entirely accurate however. Granted, humans are no longer only passive observers of planetary phenomena. As a global species our impacts are global too, and we are a distinct force *on* nature. But we are also unlike the mindless forces *of* nature. Hurricanes cannot choose to do anything but that which the laws of physics prescribe. Hurricanes are enslaved to the processes that set them in motion, whereas the actions of humans and society can be proscribed. As a society we shape our political, economic, social, and scientific institutions in ways that amplify, reduce, or modify those forces as required. Humans, unlike hurricanes, have a choice in how we exert our force.

Please do not misunderstand, this is not some Pollyanna-ish optimism as though *we* all know what needs to be done. We, the authors, are fully aware that there is no homogeneous '*we*'. Society is a patchwork of cultural, institutional, and philosophical paradigms that rarely agree and are sometimes even mutually exclusive. Yet hope dies last for good reason. In 1985, all 14 of the humpback whale (*Megaptera novaengliae*) populations around the world were greatly reduced, some by as much as 95 per cent. Then, an international moratorium on commercial whaling was introduced, and today, nine of those populations are no longer at risk, and the remaining continue to recover. Or consider the Montreal Protocol: an international treaty that came into effect in 1989 to phase out the production and use of ozone-depleting compounds such as chlorofluorocarbons. Most striking about that agreement was that global regulations were being drawn up long before scientific consensus had been established on the effect of ozone-depleting compounds. The world of the 1980s rejected the status quo, and the world of today is better for it.

Even though the challenges are different, we have a similar choice now in how we deal with invasive species, and how we act in the coming decades will be lauded or lambasted by future generations. We cannot stand behind the claim of innocent foolhardiness like our 19th- and early 20th-century forebears could, when ecology, let alone invasion ecology, was in its infancy. We know full well what causes invasive species, and we know the damage that results. Our near future is unquestionably manacled to increasing invasive species, but the long term need not be that way. And we don't need to wait for some miraculous technological breakthrough, as we already know the steps to take to change the trajectory. Our only option then is to decide, consciously, deliberately, whether to reject the status quo. We should. The future will be better for it.

References

Aagaard, K., and Lockwood, J. (2014) Exotic birds show lags in population growth. *Diversity and Distributions*, 20(5), 547–54.

Aikio, S., Duncan, R., and Hulme, P. (2010) Herbarium records identify the role of long-distance spread in the spatial distribution of alien plants in New Zealand. *Journal of Biogeography*, 37(9), 1740–51. DOI:10.1111/j.1365-2699.2010.02329.x

Bailey, S. A. (2015) An overview of thirty years of research on ballast water as a vector for aquatic invasive species to freshwater and marine environments. *Aquatic Ecosystem Health & Management*, 18(3), 261–8. <https://doi.org/10.1080/14634988.2015.1027129>

Beaury, E. M., Fusco, E. J., Allen, J. M., and Bradley, B. A. (2021) Plant regulatory lists in the United States are reactive and inconsistent. *Journal of Applied Ecology*, 58(9), 1957–66. <https://doi.org/10.1111/1365–2664.13934>

Bejder, M., Johnston, D. W., Smith, J., Friedlaender, A., and Bejder, L. (2016) Embracing conservation success of recovering humpback whale populations: evaluating the case for downlisting their conservation status in Australia. *Marine Policy*, 66, 137–41.

Bellard, C., Cassey, P., and Blackburn, T. M. (2016) Alien species as a driver of recent extinctions. *Biology letters*, 12(2), 20150623.

Bradshaw, C. J., Leroy, B., Bellard, C., Roiz, D., Albert, C., Fournier, A., Barbet-Massin, M., Salles, J. M., Simard, F., and Courchamp, F. (2016) Massive yet grossly underestimated global costs of invasive insects. *Nature communications*, 7(1), 1–8.

Cassey, P., Delean, S., Lockwood, J. L., Sadowski, J. S., and Blackburn, T. M. (2018) Dissecting the null model for biological invasions: a meta-analysis of the propagule pressure effect.

PLOS Biology, 16(4), e2005987. <https://doi.org/10.1371/journal.pbio.2005987>
Costanza, R., De Groot, R., Sutton, P., Van der Ploeg, S., Anderson, S. J., Kubiszewski, I., Farber, S., and Turner, R. K. (2014). Changes in the global value of ecosystem services. *Global environmental change*, 26, 152–8.
De Groot, R., Brander, L., Van Der Ploeg, S., Costanza, R., Bernard, F., Braat, L., Christie, M., Crossman, N., Ghermandi, A., Hein, L., and Hussain, S. (2012) Global estimates of the value of ecosystems and their services in monetary units. *Ecosystem services*, 1(1), 50–61.
Diagne, C., Leroy, B., Vaissière, A.-C., Gozlan, R. E., Roiz, D., Jarić, I., Salles, J., Bradshaw, C. J. A., and Courchamp, F. (2021) High and rising economic costs of biological invasions worldwide. *Nature*, 592(7855), 571–6. DOI: 10.1038/s41586-021-03405-6
Doherty, T. S., Glen, A. S., Nimmo, D. G. and Dickman, C. R. (2016) Invasive predators and global diversity loss. *PNAS*, 113(40), 11261–5. <https://doi.org/10.1073/pnas.1602480113>
Donovan, G. H., Butry, D. T., Michael, Y. L., Prestemon, J. P., Liebhold, A. M., Gatziolis, D., and Mao, M. Y. (2013). The relationship between trees and human health: evidence from the spread of the emerald ash borer. *American Journal of Preventive Medicine*, 44(2), 139–45.
Duffy, S. G. (1896) *My life in two hemispheres*. T. Fisher Unwin.
Gertzen, E., Familiar, O., and Leung, B. (2008) Quantifying invasion pathways: fish introductions from the aquarium trade. *Canadian Journal of Fisheries and Aquatic Sciences*, 65(7), 1265–73 <https://doi.org/10.1139/F08-056>
Green, P. T. (1997) Red crabs in rain forest on Christmas Island, Indian Ocean: activity patterns, density and biomass. *Journal of Tropical Ecology*, 13, 17–38.
Griffiths, C. J., Jones, C. G., Hansen, D. M., Puttoo, M., Tatayah, R. V., Müller, C. B. and Harris, S. (2010) The use of extant non-indigenous tortoises as a restoration tool to replace extinct ecosystem engineers. *Restoration Ecology*, 18(1), 1–7.
Grigorovich, I. A., Therriault, T. W., and MacIsaac, H. J. (2003) History of aquatic invertebrate invasions in the Caspian Sea. *Biological Invasions*, 5, 103–15.
Katsanevakis, S., Wallentinus, I., Zenetos, A., Leppäkoski, E., Çinar, M. E., Oztürk, B., Grabowski, M., Golani, D., and Cardoso, A. C. (2014) Impacts of invasive alien marine species

on ecosystem services and biodiversity: a pan-European review. *Aquatic Invasions* 9: 391–423.

Kettunen, M., Genovesi, P., Gollasch, S., Pagad, S., Starfinger, U., ten Brink, P., and Shine, C. 2009. Technical support to EU strategy on invasive species (IAS)—Assessment of the impacts of IAS in Europe and the EU (Final draft report for the European Commission). Institute for European Environmental Policy (IEEP), Brussels, Belgium)

Le Corre, M., Danckwerts, D. K., Ringler, D., Bastien, M., Orlowski, S., Morey Rubio, C., Pinaud, D., and Micol, T. (2015) Seabird recovery and vegetation dynamics after Norway rat eradication at Tromelin Island, western Indian Ocean. *Biological Conservation*, 185, 85–94. <https://doi.org/10.1016/j.biocon.2014.12.015>

Le Maitre, D. C., Gaertner, M., Marchante, E., Ens, E., Holmes, P. M., Pauchard, A., O'Farrell, P. J., Rogers, A. M., Blanchard, R., Blignaut, J., and Richardson, D. M. (2011) Impacts of invasive Australian acacias: implications for management and restoration. *Diversity and Distributions*, 17(5), 1015–29. <https://doi.org/10.1111/j.1472-4642.2011.00816.x>

Le Page, S. L., Livermore, R. A., Cooper, D. W., and Taylor, A. C. (2001) Genetic analysis of a documented population bottleneck: introduced Bennett's wallabies (*Macropus rufogriseus rufogriseus*) in New Zealand. *Molecular Ecology*, 9(6), 753–63. <https://doi.org/10.1046/j.1365-294x.2000.00922.x>

Lewis, S. L., and Maslin, M. A. (2015) Defining the Anthropocene. *Nature*, 519, 171–80.

Li, X., Holmes, T. P., Boyle, K. J., Crocker, E. V., and Nelson, C. D. (2019) Hedonic analysis of forest pest invasion: the case of emerald ash borer. *Forests*, 10(9), 820. <https://doi.org/10.3390/f10090820>

Lockwood, J. L., Cassey, P., and Blackburn, T. (2005) The role of propagule pressure in explaining species invasions. *Trends in Ecology & Evolution*, 20(5), 223–8. <https://doi.org/10.1016/j.tree.2005.02.004>

Morton, O. (2016) *The planet remade*. Princeton University Press, p. 220.

Mitchell, C. E. and Power, A. G. (2003) Release of invasive plants from fungal and viral pathogens. *Nature*, 421, 625–7.

Murray, C. C., Pakhomov, E. A., and Therriault, T. W. (2011) Recreational boating: a large unregulated vector transporting marine invasive species. *Diversity and Distributions*, 17(6), 1161–72. <https://doi.org/10.1111/j.1472-4642.2011.00798.x>

Ngorima, A., and Shackleton, C. (2019) Livelihood benefits and costs from an invasive alien tree (*Acacia dealbata*) to rural communities in the Eastern Cape, South Africa. *Journal of Environmental Management*, 229, 158–65. <https://doi.org/10.1016/j.jenvman.2018.05.077>

Nyoka, B. I. (2003) *Biosecurity in forestry: a case study on the study of invasive forest tree species in Southern Africa*. Forest Biosecurity Working Paper FBS/1E. Forestry Department. FAO, Rome.

Palmer, T. S. (1894) *The danger of introducing noxious animals and birds*. US Department of Agriculture.

Pimentel, D. (ed.) (2011) *Biological invasions: economic and environmental costs of alien plant, animal, and microbe species*. CRC Press.

Seebens, H. (2021). Alien Species First Records Database (Version 2) [Data set]. Zenodo. <http://doi.org/10.5281/zenodo.4632335>

Shackleton, R. T., Shackleton, C. M., and Kull, C. A. (2019) The role of invasive alien species in shaping local livelihoods and human well-being: a review. *Journal of Environmental Management*, 229, 145–57.

Siegert, N. W., McCullough, D. G., Liebhold, A. M., and Telewski, F. W. (2014) Dendrochronological reconstruction of the epicentre and early spread of emerald ash borer in North America. *Diversity and Distributions*, 20(7), 847–58. <https://doi.org/10.1111/ddi.12212>

Stiling, P. and Cornelissen, T. (2005) What makes a successful biocontrol agent? A meta-analysis of biological control agent performance. *Biological Control*, 34(3), 236–46. <https://doi.org/10.1016/j.biocontrol.2005.02.017>

Tobin, P. C., Kean, J. M., Suckling, D. M., McCullough, D. G., Herms, D. A., and Stringer, L. D. (2014) Determinants of successful arthropod eradication programs. *Biological Invasions*, 16, 401–14.

Turbelin, A. J., Diagne, C., Hudgins, E. J., Moodley, D., Kourantidou, M., Novoa, A., Haubrock, P. J., Bernery, C., Gozlan, R. E., Francis, R. A. and Courchamp, F. (2022) Introduction pathways of economically costly invasive alien species. *Biological Invasions*, 24, 2061–79. <https://doi.org/10.1007/s10530-022-02796-5>

Valentin, R. E., Fonseca, D. M., Gable, S., Kyle, K. E., Hamilton, G. C., Nielson, A. L., and Lockwood, J. L. (2020) Moving eDNA surveys onto land: strategies for active eDNA aggregation to detect invasive forest insects. *Molecular Ecology Resources*, 20(3), 746–55. <https://doi.org/10.1111/1755-0998.13151>

Walsh, J. R., Carpenter, S. R., and Vander Zanden, M. J. (2016) Invasive species triggers a massive loss of ecosystem services through a trophic cascade. *Proceedings of the National Academy of Sciences*, 113(15), 4081–5.

Wilson, J. R. U., Dormontt, E. E., Prentis, P. J., Lowe, A. J., and Richardson, D. M. (2009). Something in the way you move: dispersal pathways affect invasion success. *Trends in Ecology & Evolution*, 24(3), 136–44. <https://doi.org/10.1016/j.tree.2008.10.007>

Further reading

Invasive species, general

Burdick, A. (2006) *Out of Eden: an odyssey of ecological invasion.* Farrar, Straus and Giroux.

Elton, C. S. (2020) *The ecology of invasion by plants and animals,* 2nd edition. Springer.

Lockwood, J. L., Hoopes, M. F., and Marchetti, M. P. (eds) (2013) *Invasion ecology,* 2nd edition. Wiley-Blackwell.

Pimentel, D. (ed.) (2011) *Biological invasions: economic and environmental costs of alien plant, animal, and microbe species.* CRC Press.

Rotherham, I. D. and Lambert, R. A. (eds) (2013) *Invasive and introduced plants and animals: human perceptions, attitudes and approaches to management.* Routledge.

Simberloff, D. (2013) *Invasive species: what everyone needs to know.* Oxford University Press.

Invasive species in particular ecosystems

Low, T. (2002) *Feral future: the untold story of Australia's exotic invaders.* University of Chicago Press.

Meinesz, A. (1999) *Killer algae.* University of Chicago Press.

Perez, L. (2012) *Snake in the grass: an Everglades invasion.* Pineapple Press.

Rapai, W. (2016) *Lake invaders: invasive species and the battle for the future of the Great Lakes.* Wayne State University Press.

Managing invasive species

Carlton, J. and Ruiz, G. M. (eds) (2003) *Invasive species: vectors and management strategies*. Island Press.

Robinson, A. P., Walshe, T., Burgman, M. A., and Nunn, M. (2017) *Invasive species: risk assessment and management*. Cambridge University Press.

Saunders, J. (2016) *Invasive species management: control options, congressional issues and major laws*. NOVA Science Pub, Inc.

Ecology

Bowman, W. D. and Hacker, S. D. (2021) *Ecology*, 5th edition. Oxford University Press.

Marchetti, M. P., Lockwood, J. L., and Hoopes, M. F. (2023) *Ecology in a changing world*. W. W. Norton & Company.

Environmental philosophy

Bassham, G. (2020) *Environmental ethics: the central issues*. Hackett Publishing Company, Inc.

Crosby, A. (2015) *Ecological imperialism*, 2nd edition. Cambridge University Press.

Frawley, J. and McCalman, I. (eds) (2014) *Rethinking invasion ecologies from the environmental humanities*. Routledge.

Jamieson, D. (ed.) (2003) *A companion to environmental philosophy*. Blackwell Publishing.

Conservation

Diehm, C. (2020) *Connection to nature, deep ecology, and conservation social science: human–nature bonding and protecting the natural world*. Lexington Books.

Kareiva, P. and Marvier, M. (2017) *Conservation science: balancing the needs of people and nature*, 2nd edition. W. H. Freeman.

Index

For the benefit of digital users, indexed terms that span two pages (e.g., 52–53) may, on occasion, appear on only one of those pages.

A

B

C

D

E

F

G

H

I